Geoffrey Williams drives Sir Thomas Lipton *crisply through a quiet sea. A ruthless approach to the Race together with the classic lines drawn by the designer Robert Clark brought man and boat first to Newport after a record crossing of the Atlantic.*

THE *OBSERVER* SINGLEHANDED
TRANSATLANTIC RACE

ATLANTIC VENTURE

John Groser

WARD LOCK & CO. LTD
London and Sydney

7063 1034 9

Cox and Wyman Limited
London, Fakenham and Reading

Contents

Foreword

This is a book about the Singlehanded Transatlantic Race sponsored by the London *Observer*, and in particular about the third race of the series, held in 1968, which stirred up a good deal of comment and criticism.

John Groser of *The Observer* is the best possible person to write about it, having been closely involved with sponsors, organisers, competitors, and local public opinion both at Plymouth before the start and at Newport, R.I., after the finish.

Seen in perspective, and fanning away the clouds of publicity that seem to surround it, the race is nothing more than a small specialised sideshoot growing out of the great body of recreational sailing. Only a minority of sailing people have any inclination to sail in it, but these tend to get strongly obsessed. Some of those who don't want to sail in it can often be heard raising objections to the race, and to the way in which it is organised.

In spite of pressures from commercial and national interests, it remains a sporting event organised by a committee of amateur yachtsmen for the benefit of a small group of other yachtsmen. Technically, it performs a useful service in subjecting the conflicting claims of yacht designers, yacht salesmen, and specialist yachting groups to the unbiased judgement of the sea itself. From a technical viewpoint, failures are at least as important as successes, and the results of the race should help anyone who is looking for the truth about the performance of sailing-boats in open water.

A huge part of its success is due to the Royal Western Yacht Club of England, whose Race Committee, led by Jack

Odling-Smee, tackles its job with speed and efficiency, and in calm disregard of the brickbats being flung at it by any number of non-participants.

I think it is wrong to regard the race as being an heroic test of courage and endurance. These qualities are only needed to a marked degree if there is something wrong with the boat's design or condition, or with the ability or state of health of its crew. Very little progress has yet been made in designing boats that are not only large and fast but also *easy to handle*, and any designer with his eye on future Single-handed Transatlantic Races might well start by tackling this side of the problem more seriously.

Curdridge, September 1968   H. G. HASLER

Preface

For six months this year I was involved, one way and another, with the Singlehanded Transatlantic Race. In that time I was privileged to meet the competitors, some of whom I came to know well. They all impressed me with their love, or perhaps I should say joy, of the sea. Yachtsmen in the race may not care to hear it described as a battle between man and the elements, and probably they do not see it as such. But for the layman, who is able to sit in an air-conditioned bar in Newport, waiting for the boats to come in, the eloquence of a competitor's log speaks for itself.

This book has no pretensions to be the history of the three races, nor has it an official line about sponsorship or any of the other controversial issues that made this year's race such an absorbing event of international interest. It is simply a personal record of my association with thirty-five quite extraordinary people, some of whom I felt were changed by three thousand miles of Atlantic, and all of whom affected me deeply.

While writing the book, I relied heavily on the assistance and encouragement of Anthony Churchill, and without his help it could never have been completed in the time at my disposal. For his technical advice I am deeply grateful. He also read the manuscript and made many constructive suggestions. He was entirely responsible for organising the illustrations for the book and introduced me to Christopher and Julian Everitt who produced the scale drawings of all the yachts.

The maps and charts were drawn by Roger Humm, who laboured long to ensure that they were accurate.

9

My thanks are due to Elizabeth Balcon and her assistants, without whose research I should have been lost.

For facilitating my use of Race Committee files, I am grateful to Terence Shaw and Jack Odling-Smee, who gave me every assistance with the historical part of the Introduction, as did Blondie Hasler.

Acknowledgement is also due, and is gladly made, to: *Yachting World* for the basis of the statistical table that comprises part of Appendix B, and for permission to reproduce extracts from an editorial article. *Yachting and Boating* for the idea and basis of the four weather charts. The Meteorological Office (Crown Copyright: H.M. Stationery Office) for allowing the use of their weather charts for the month of June. For the excellent photographs I have to thank Beken & Son, David Baker (courtesy of Perspective Publications), Pierre Fouquin, Photo Lanoue-Bateaux, Colin Rowe, Reijo Rüster, Roger Smith (courtesy of *Yatching and Boating Weekly*) and Chris Smith (courtesy of *The Observer*).

Richmond, September 1968   JOHN GROSER

I

Introduction

Thirty-four men and one woman bore eloquent testimony to the timeless challenge of the ocean as they milled about the restricted starting area in Plymouth Sound on the first day of last June. They were the contestants in the third *Observer* Singlehanded Transatlantic Race, and had been variously heralded as brilliant sailors, national heroes, cranks, solitaries and misfits.

With hardly a breath of wind to blow them out to sea, the boats bobbed slowly about. Eric Tabarly was one of the first across the start line in his monstrous silver-grey aluminium trimaran, *Pen Duick IV*. He himself had described the boat as a 'racing machine' and she more nearly resembled a Bailey bridge than a yacht, with great struts of metal joining floats to hull and a wide semicircular steel track for the mainsail. She looked enormous, invincible and very fast.

Slightly astern, and completely dwarfed by the trimaran, came *Cheers* with Tom Follett working furiously to give her some way. The brilliant yellow proa was a complete contrast to Tabarly's metal machine. *Cheers* seemed delicate, a creation conjured up by Salvador Dali, with just a hint of a Leonardo da Vinci flying machine and an obvious Martian influence. But for all that she was pretty in a sculptured sort of way – and she also looked threateningly fast.

The more conventional yachts were strung out in the Sound. The incredibly beautiful *Sir Thomas Lipton* seemed just like a toy yacht on some gargantuan pond. Pale blue and white and Geoffrey Williams had all her sails set, though the wind hardly filled them.

Stephen Pakenham was rowing *Rob Roy*, and a rather damp wit on the Press boat shouted out that Oxford were leading by five lengths. Bernard Waquet knelt purposefully in the bow of *Tamoure* and paddled kayak-style. Edith Baumann sailed into the breakwater.

To many of the onlookers that wretchedly wet Saturday morning, some of the entrants must have indeed appeared foolhardy, for their boats looked puny, frail and insignificant specks on the grey and intimidating expanse of water. Could one man in a dainty twenty-foot sloop pit his craft against the terror of the Atlantic? Would many of the boats in fact reach Newport, Rhode Island, where the race was supposed to end.?

A smiling, gentle man watched the small flotilla set out. H. G. 'Blondie' Hasler, a retired lieutenant-colonel in the Royal Marines, felt confident of the answers to those questions. He himself had sailed a strange and enchanting boat called *Jester* in both the previous races. *Jester*, with her new owner aboard, was out there now, disappearing imperceptibly into the mist of rain with the other yachts. Hasler may perhaps have wondered what he had started in devising such an adventure. But he knew and loved the sea and was fascinated by it – and he supposed that all the lone sailors were too. Also, he felt at home in small boats.

In 1941, an already balding Marine captain had suggested that small boats be used to destroy shipping in enemy harbours. A wild idea, but Captain Hasler seemed to be able to get away with wild schemes, and by the early summer of 1942, Operation Frankton was under way. Hasler, who was appointed leader of the expedition he had devised, selected thirty marines who wanted to have a personal go at the enemy, no matter what the danger. From this group he eventually chose nine: with him, they were to come to be known as the 'Cockleshell Heroes'. Their daring raid on the ships in the German-held harbour at Bordeaux is now part of history. Six enemy vessels were sunk, and Hasler lost eight of his men. The operation was accredited a complete success.

Before the War even, and before his daring feat of seamanship in a canoe, Hasler had been called by the sea. As a subaltern he had sailed an open dinghy singlehanded from Plymouth to Portsmouth, and back – an escapade which nowadays might command little attention in the popular Press, but which then was certainly a nautical event years ahead of its time.

Until 1948, when he left the Marines, Hasler seems to have been satisfied with what he calls 'orthodox' sailing.

'Certainly I had done unorthodox things' – Hasler is the master of understatement – 'but after 1948, I became obsessed with the idea of experimenting. With boats, with everything really.' He began to devote himself to a full-time career in unorthodoxy. He wanted to invent and develop things, and he did so with some phenomenal results.

Basically he became tired of sailing big boats that required a crew. He found that the problem of a crew was that he needed at least one other person who wanted, and was free, to sail a particular yacht at a particular time and place. So he turned to smaller boats, and discovered in singlehanded sailing much of the solitary excitement he had been looking for. Talking about this period of his life now, so many years later, Hasler has a faraway look, the look of a visionary perhaps. I saw that same look in the eyes of each of the competitors as he stepped ashore in Newport at the end of the race this summer.

If solo sailing was to be made enjoyable, Hasler needed a really efficient, properly designed yacht. In the winter of 1952–53 he built *Jester*. The rig and layout of this twenty-five foot Folkboat have been called the most fundamental advance in yacht design for fifty years, and Hasler experimented with her all the time. His aim has been to evolve a small seaworthy boat suitable for sailing easily – 'I am a very lazy man' – with a one man crew.

For an innovator and initiator, a developer and designer, there is no great satisfaction in untested or untried achievement or apparent success. With motor cars, the greatest stimulant to development is competitive racing. So, Hasler thought, with yacht design. The idea of a great solo race across the Atlantic came to him in 1956. Partly he wanted to see if he had really been successful with his own designs, and partly he hoped to find kindred spirits with the same flair for the unorthodox, the same lust for adventure, and perhaps (dinghies and cockleshells) with the same need of the drug called danger.

'I wanted to have a race without handicapping formulae, the only limiting factor being what one man could do on his own. The minimum of rules – I don't like rules very much – and in any case I was having a reaction against R.O.R.C. racing. I'd done a lot of that. I suppose it's a form of conceit, but I tend to be agin the establishment and like things to be as unorthodox as possible.' He obviously also likes things to be difficult, which is why he chose the East to West crossing as the course for the race.

In March 1959, Hasler found a kindred spirit. David Lewis, a doctor of medicine who had spent a large part of his childhood in New Zealand, wrote to Hasler saying that he was interested in the idea of the race and that he thought he would like to compete.

Hasler soon found a second spirit when Francis Chichester wrote to him in October. Chichester was also a pioneer and inventor, an unorthodox man already famous for his exploits in the air. But in 1959 he had only been sailing some seven years and was probably not nearly so fine a sailor as the other two.

In January 1957, Hasler had approached *The Observer* to see if the newspaper would sponsor the race, and although the

*The sharp prows of Eric Tabarly's trimaran, Pen Duick
IV, never dug deep into the Atlantic. Strikes in
France delayed her launching, and the 1964 winner had
no time to sail her through her teething troubles. A
collision and damaged gear led to an early retirement.*

The start at Plymouth in the rain and a ghosting wind on the 1st June gave little warning of the coming storms of the 11th and 12th, the fiercest many of these hardened yachtsmen had ever sailed through.

management were favourably disposed, they decided by the end of the month that they could not in fact do so. Hasler thought of that newspaper because he knew the Editor, David Astor (they had met in the Marines during the War), and he thought that the paper's liberal, slightly unorthodox traditions would enable the policy makers within the organisation to look favourably on a more than slightly unorthodox venture. But, initially, he was rebuffed. So in February, he wrote to the Slocum Society in New York. The society was founded in memory of Joshua Slocum, the nineteenth-century adventurer who sailed his yacht *Spray* singlehanded round the world from 1895 to 1898. For obvious reasons, Hasler hoped that the society might back his idea. Gordon Mc-Closky, the society's honorary secretary wrote back to say that he liked the scheme, and that he would take soundings among the senior members to see what their reactions would be.

Much as Hasler hated them, he realised that there would have to be some sort of rules, and he sent a draft to McClosky, who replied in May that on the whole the general feeling of the members of the society was favourable, but that it would not be possible to organise a race until 1960. Hasler had proposed July 1958 as the starting date, but he accepted the postponement, and started looking for an English yacht club to organise things at this end and also to be responsible for the start. Printed rules, the original draft with slight amendments and alterations, were published by the Slocum Society in August 1957, and a Press release was sent out giving the news and information about the proposed race.

There was now a major hitch, in that Hasler had some difficulty finding an organising yacht club in England. The Island Sailing Club refused to take any part, and in October Hasler approached F. G. 'Tiny' Mitchell and the Royal Corinthian Y.C. By the end of the month, Mitchell had agreed to handle the start from Cowes, but said that the club would not take any part in organising or controlling the race.

From this time onwards, the Slocum Society seemed to become increasingly anxious about the possible danger of such a race. They had received numerous dissenting opinions from all sorts of people, many of them established and experienced ocean sailors. They began to retreat, trying to change the rules and alter the spirit of Hasler's idea. He told them that if they did not abide by the published rules, he would have to look elsewhere for another sponsor.

Without a suitable sponsor there were likely to be problems in staging the event at all. And Hasler next approached the *Daily Express*. In July, he got a negative answer from them. But almost immediately after that, he heard from Christopher Brasher, who was then the Sports Editor of *The Observer*, and who had apparently persuaded the manager of the newspaper to offer a trophy for the race.

Hasler has strong views about the part Brasher played in getting the race to sea. Brasher was then a fairly blustery thirty-year old, who had capped his magnificent run with Bannister and Chataway at Iffley Road by winning a steeplechase gold medal in the Melbourne Olympics of 1956. He had left industry to be Sports Editor of the newspaper, and he also was known to be another slightly unorthodox figure. Had he been a sailor, I am sure he would have wanted to take part in the transatlantic race. As it was, he found in sports such as orienteering much of the satisfaction the others found in solo sailing.

Brasher and Hasler now discussed the possibility of greater participation on the part of the paper, and the whole project began to gather a new momentum. Lewis went to see Hasler to tell him that he had qualified to take part in the race, and that he was ready to go.

It was as well that *The Observer* had begun to take an interest in the race, for in October 1959 John Pflieger, the Com-

Bruce Dalling sailed Voortrekker *from Cape Town to Plymouth for the Race. A Springbok on her bows, she slips along with spinnaker, main and mizzen staysail set.*

Goodwin II *is sistership to* Fione, *and at 19 feet they were the shortest yachts in the Race. Ake Mattsson sailed into Newport after fifty days at sea. He was disqualified for taking on supplies during the crossing.*

modore of the Slocum Society (Pflieger was tragically lost at sea in 1966 while sailing singlehanded from Bermuda to Antigua) wrote to Hasler to say that the society had decided not to hold a race, certainly not of the sort proposed. Instead, they intended to hold a 'cruising competition'. Competitors could start any time they liked in 1960, and could use their engines if they so wished. Printed rules for this competition were published in November, and included the words 'the Slocum Society is not going to sponsor a race'. That circular was never withdrawn nor amended.

Meanwhile, Chichester had written to Hasler about the society's volte-face, and Hasler subsequently circulated the British yachting Press, advising them to ignore any material or information about the event that did not emanate from him.

In October, he again asked Mitchell of the Royal Corinthian Y.C. if the club would organise as well as start the race, and again they regretfully declined. Then Chichester sent out a Press release explaining that Hasler and he were organising the British end of the race, and that the Slocum Society, notwithstanding their refusal of sponsorship, would handle the finish. Hasler wrote to Pflieger saying that he had re-assumed responsibility for organising, publicising and controlling the race, though he hoped to work in harmony with the society, especially at the finish – which was to be in New York.

In early December, Hasler and Chichester approached the Royal Western Y.C. of England (known to many as the cradle and nursery of ocean racing and sailing) to see if they would be responsible for organising the race. In retrospect, Hasler says that he cannot imagine why he did not go to the R.W.Y.C. in the first place. It was obvious that, even though *The Observer* would be sponsors, the race needed an independent organiser if it were to be run really efficiently. A newspaper cannot organise a race of this sort, nor really can the competitors – the more so once they are under way.

Hasler emphasises that he, Chichester, Lewis and Val Howells had all met and decided to race whether an organiser could be found or not, and Chichester's remark that he and Hasler would race for a half-crown wager is now so well publicised as to be part of authorised yachting folk-lore. The fourth of the original entrants for the race was a Welsh farmer. Howells had been in the Merchant Navy, and was a considerable sailor. He got to know Eric Tabarly quite well before and after the 1964 race, and the Frenchman expressed the opinion that Howells was a fine yachtsman, '. . . remarkably quick at handling a boat, and with great strength and energy'.

During the month of December 1959, the Royal Western Y.C. discussed the problems of organising the event, and although some fears and doubts were voiced about the possible public outcry should there be any mishaps during the race, it was decided that the club would take full responsibility for the organisation. George Everitt, then the secretary of the R.W.Y.C., wrote to Hasler to give him the good news on 31 December. Just before Christmas, Brasher had contacted Hasler to say that *The Observer* was now prepared 'to jump in pretty deeply' – in other words that the newspaper would accept the full responsibilities and obligations as official, and sole, sponsors.

The part played by the Royal Western Y.C. is extremely important. It would be ludicrous to say that without the club there could have been no race – Hasler and the other three had already said that they would make the crossing anyway. But without the R.W.Y.C. I doubt that the race would either have been so outstandingly successful, or whether it would have grown in the way it has.

The club set up a Race Committee under the chairmanship of Jack Odling-Smee (a retired lieutenant-colonel in the Army) who was Rear-Commodore of the club. The Commodore was Sir Winston Churchill. Odling-Smee had seen straight away the brilliant vision behind Hasler's scheme, and had offered his wholehearted support. He admits that there was no little opposition at times, but as chairman he was able to take his committee along with him, and preparations for the first race proceeded apace.

There were eight entries for the race; *Jester* (Blondie Hasler), *Gipsy Moth III* (Francis Chichester), *Cardinal Vertue* (David Lewis), *Eire* (Val Howells), *Cap Horn* (Jean Lacombe), and three who did not start – the American Arthur Piver, an Englishman named Mike Ellison, and Karminski, who sent in an application from Hamburg, but who never arrived at Plymouth.

FINAL PLACINGS IN THE 1960 RACE

			Arrival	
Place	Yacht	Crew	Date	Days
1	Gipsy Moth III	F. Chichester	22/7	40
2	Jester	H. G. Hasler	30/7	48
3	Cardinal Vertue	D. Lewis	6/8	56
4	Eira	V. Howells	13/8	63
5	Cap Horn	J. Lacombe	24/8	74

Cardinal Vertue put back to Plymouth with a damaged mast and set sail again on 13 June.
Eira had to put into Bermuda after being damaged in a gale.
Cap Horn was taken in tow for a few hours because of high seas.

At 10 a.m. on Saturday 11 June, Hasler, Lewis, Howells and Chichester (in that order) sailed across the starting line in Plymouth Sound to begin 'the great race of the century'. Lacombe had only arrived on the Friday evening after a rough crossing from Le Havre, and did not set out in pursuit of the other four until 14 June. Forty days later, Chichester sailed into New York Harbour after a remarkable passage, in a time which many experts thought 'fantastic' and which easily broke the previous record time for a singlehanded crossing.

Hasler came second, followed by Lewis and Howells (who had to put into Bermuda for repairs). Lacombe struggled into port after sixty-nine days in the tiny eighteen-foot boat he was charged to deliver to her new owner in New York.

The race had been an outstanding success, and had been followed with great interest by yachting enthusiasts. There had been no serious accidents, despite some appalling weather in mid-Atlantic, and the Cassandras had been proved wrong in their dire prognostications.

The actual finishing line was off the Ambrose Lightship, at the southernmost tip of Long Island, and the finish was organised jointly by the Slocum Society and the Sheep's Head Bay Y.C., who were apparently tremendously hospitable and charming. But they were a small club, without the extensive facilities of some of their bigger American brethren; and there was also the problem of finishing a race of this sort at the entrance to such a busy port.

Odling-Smee began to wonder if the race should terminate at some other East Coast city. He happened to meet in London a member of the New York Y.C. who suggested that a good place to finish might be Newport, Rhode Island. On the instructions of the Earl Morley, then Vice Commodore of the Royal Western, Odling-Smee's Race Committee wrote to the New York Y.C. to ask their advice. They, in turn, referred the committee on to the Ida Lewis Y.C. in Newport, and so began the association that, happily, still continues.

The Ida Lewis is a rather imposing club, built on a small island at the end of a long spit that juts into Newport Harbour. It is as exclusive as any yacht club can be, and perhaps because of its geographical location and its long, vaguely intimidating causeway, seems somewhat remote from the everyday affairs of Newport.

The 1964 race finished in Newport, and proved to be an even greater success than its predecessor. Hasler had written in 1960 'Above everything else, I want this race to be a success, to become a regular and recognised event. I believe that the design of small seagoing sailing-boats is still in its infancy, and that this is the best way of encouraging it to grow up.'

Hasler's idea had become by 1964 a regular and recognised event, and to prove the truth of his belief, there were fifteen starters, fourteen of whom set out from Plymouth on Saturday 23 May. In addition to the five sailors who had taken part in the 1960 race, and who were all competing again, there was the Dane, Axel Pedersen, who only arrived in Plymouth the morning the race started, and who set out three days later in his little ketch *Marco Polo*. Leading the other fourteen out into the Channel was a quiet, bull-necked French naval lieutenant called Eric Tabarly. His ketch *Pen Duick II* was a formidable looking boat – perhaps too big for one man to handle. In any case, it was thought, Chichester would win again. How wrong the thinkers were to be proven less than a month later when the Frenchman sailed into Newport in the thoroughly impossible time of twenty-seven days.

Chichester came second, improving on the passage he sailed in 1962 when he had made another solo crossing in an effort to break his 1960 race record. Hasler again did well, bettering his 1960 time by ten days; and the multihulls, of which there were three in the race, did rather better than much of the adverse criticism at the time might have led the unknowing to think. At least *Rehu Moana*, *Misty Miller* and *Folatre* did no worse, and were not much slower, than good monohulled boats of comparable size.

Piver, who had entered the 1960 race and had not arrived

FINAL PLACINGS IN THE 1964 RACE

Place	Yacht	Crew	Date	Arrival D.	H.	M.	Handicap D.	H.	Place
1	Pen Duick II	E. Tabarly	19/6	27	03	56	21	23	1
2	Gipsy Moth III	F. Chichester	22/6	29	23	57	22	18	2
3	Akka	V. Howells	25/6	32	18	08	24	07	3
4	Lively Lady	A. Rose	29/6	36	17	30	27	09	5
5	Jester	H. G. Hasler	30/6	37	22	05	25	04	4
6	Stardrift	W. Howell	30/6	38	03	23	27	12	6
7	Rehu Moana	D. Lewis	30/6	38	12	04			*
8	Illala	R. M. Ellison	8/7	46	06	26	34	20	9
9	Golif	J. Lacombe	8/7	46	07	05	30	00	7
10	Vanda Caelea	R. Bunker	12/7	49	18	45	32	22	8
11	Misty Miller	C. M. Butterfield	15/7	53	00	05			*
12	Ericht	G. Chaffey	22/7	60	11	15	42	23	10
13	Folatre	D. Kelsall	24/7	61	14	04			*
14	Marco Polo	A. Pedersen	26/7	63	13	30	44	21	11
	Tammie Norie	R. McCurdy	retired						

*Multihulled boats were not eligible for the Handicap Trophy.
Akka was delayed at the start after being rammed by a pleasure boat, and put into Southern Ireland for repairs.
Misty Miller and *Ericht* put into the Azores for repairs.
Marco Polo did not leave Plymouth until 26 May.
Folatre returned damaged to Plymouth and put out again on 19 June.

for the start, entered again in 1964. Much to his disappointment his trimaran *Bird* was damaged, as his 1960 boat had been, and he was still under repairs in Bermuda on 23 May when the others left Plymouth.

Another American who could not take part although he had submitted his entry was Charles McLendon in *Morna*. Three weeks before the race, while the boat was moored in Yarmouth harbour, there was an explosion on board, and the damage was too great to be repaired in time for the start.

Tabarly, who, contrary to many of the reports that appeared, was not 'totally unknown' before his victory (he was in fact well known as an offshore racer of considerable skill and success) became a hero overnight. He was made a Chevalier of the Legion of Honour by a jubilant de Gaulle, and was regarded by the French for his victory over the sea (and the English) rather as they have come to regard Jean-Claude Killy for his victories on the snow slopes.

There had been an innovation for the 1964 race. A system of handicapping was introduced whereby the smaller boats would be given a chance of 'winning' the race on corrected time, and also a special Handicap Trophy. Tabarly wrote in his book *Lonely Victory* that he was never in favour of having handicap placings, and he thought that some competitors might be encouraged to enter small boats. It is difficult to agree with him entirely, as the object of the system is to encourage precisely those entrants who have neither the inclination nor the financial resources to sail a large boat in the race.

In 1960, Hasler had insisted that there should be no handicaps because of the impossibility of rating new and unconventional designs. That argument can be understood, and the Race Committee allowed for it in 1964 when they determined that only monohulls should be eligible for the handicap prize.

The last man home, a remarkably cheerful Pedersen, arrived long after Tabarly had left for France. But at least there had been no shipwrecks, and every boat (except for the one that retired) in a large fleet – fifteen competitors were thought to constitute a temptingly hazardous field – reached Newport.

It was in Rhode Island that many of the problems of the race arose. To begin with, Tabarly finished much sooner than had been thought possible, and very nearly became entangled with the enormous Bermuda Race fleet which was about to set out from Newport on the final race of the Onion Patch series.

Then it was claimed in Newport that no one there had been told that there was a transatlantic race in progress, and that the city had not had adequate time to prepare a reception for the arriving yachtsmen.

Hasty, and in the circumstances more than generous welcoming arrangements, came to an abrupt end long before the latecomers sailed into Newport, and it was said that there was no one there to look after them.

Somewhere along the line there must have been a breakdown in communications, and Odling-Smee's committee decided – rather than holding lengthy and acrimonious post-mortems – to ensure that the finish in 1968 should be organised as smoothly as the start.

The Royal Western Y.C., who had already in 1964 received inquiries about the next race, and *The Observer*, realised that they had a major sporting event on their hands. Hasler's brainchild had grown into a very big boy indeed. When the boy came to play the game next in 1968, he was to find that he had become extremely popular – but he, and his parents, were also to find that they would be severely censured for being callous and careless of human life, and some things much worse.

2

Preparations

No sooner had he won the 1964 race, than Tabarly was thinking about, and making preparations for, the next one. Four others of the competitors in that race had also given notice that they intended to sail in 1968. Tabarly made the vital observation, which for various reasons he was unable to follow four years later, that preparations for the race needed to be carefully organised, detailed and above all methodical.

For ocean racing without a crew, the speed of the boat and its ease of handling are of the greatest importance. Tabarly adds that comfort is essential as well, though I was to see little evidence of this in many of the boats that took part this year. *Spirit of Cutty Sark*, the production boat that anyone can buy, was beautifully fitted-out – she even had a carpet sweeper on board with which to dust the bulkhead to bulkhead carpeting – and might have been called comfortable. But then how comfortable can one man on his own be in mid-Atlantic? Most of the competitors spoke of being continuously wet, and having to sleep in sodden bunks.

The major obstacle to most would-be entrants in the 1968 race was obviously money. The cost of a fast, easy-to-handle racer can be considerable; the more so if the yacht must also be long at the waterline – which, as will be seen, and as Tabarly had already suggested, was to be the winning formula in the third race.

Not all the competitors were to look for sponsors to back them; in fact very few did, and even fewer had any success in their quest. Geoffrey Williams wrote over three thousand letters asking for sponsorship, before Allied Suppliers agreed to put up most of the money for his challenge. It must be emphasised that because competitors were sponsored, it did not mean that they were making any financial gain out of the project. All the entrants in the race, whether they owned their boats at the end or not, had to invest personal savings in the venture. They had to sacrifice their savings, sometimes their jobs, and in acute cases of race fever, even their families.

Williams came back from America in 1967 with a large savings account (he says about £3,000) from which to launch the *Lipton* enterprise, and he described the fitting out of a boat as being like equipping a woman's wardrobe, in that it is never complete. 'It costs a great deal of money. . . .'

Quite apart from winning the race (which also proved how carefully organised and methodical his preparations were) Williams must also have been outright winner of the perseverance stakes. He realised that there was no point in allowing an expensive challenge to go off half-cocked, and was tireless in his appeals for help to commerce and industry. Because so much criticism has been levelled at the winner for the 'commercialism' of his enterprise, I think it is vital to stress that he put considerably more effort (and not just in sailing the boat) into the whole venture than did any of his sponsors – who in the end gained far more out of the victory than did Williams, and more than they could have imagined.

Finding a sponsor was no easy task, as the endless letter-writing shows, and it is a misconception if anyone thinks that would-be backers came rushing to prospective competitors with offers of help. Williams said to me after the race, talking about his search for sponsorship: 'It was a bloody hard slog.'

The actual costs of the boats varied, from David Pyle's home-made, do-it-yourself ketch *Atlantis III*, the hull of which cost him around £500 to build (though a great deal more in time and effort), to the much-quoted and I think probably rather conservative estimate of £35,000 for *Pen Duick IV*.

Many of the boats were not built specially for the race, and although Noel Bevan already owned the gracious old ocean racer *Myth of Malham*, the cost of fitting her and getting her into the race was enormous. It would be difficult to put a figure on it, but Bevan's wife explained to me just how great the deprivations were, and how extensive the sacrifices. The Bevans sold their house so that he could compete.

National pride was greatly involved, and nowhere more so than in France. Ocean racing is a sport that interests very few Frenchmen as yet. But because Tabarly has been winning the toughest international races for France in French designed and built boats, he has become a national sporting hero, and his Government is now anxious to promote sailing. From 1964 onwards, Tabarly began to find himself more and more associated with official sport.

Because of this association, Tabarly is worried about the talk of Government sponsorship.

'Everyone thinks that the Government helps to pay for my boats. This is not true. *Pen Duick III* and *IV* are completely owned by me.' Nevertheless, he bows to commercial pressures when it comes to financing his boats and his ocean racing. His 1964 book, together with the proceeds from the sale to a naval training school of *Pen Duick II*, helped pay for *Pen Duick III*. Contracts with magazines and commercial radio have paid for *Pen Duick IV*, though because he did not finish the race, much of the money advanced for the trimaran must be paid back by him, and he is at the time of writing still heavily in debt.

In a year when South Africa was banned from taking part in the Olympic Games, Bruce Dalling represented his country's main hope of acquiring international sporting prestige. He was clearly worried by the burden.

'My father writes that enthusiasm at home is fantastic. I'm afraid that people there have no idea how fierce the

competition is. I hope they're aware that there are ten or
twelve boats just as big.'

Dalling certainly has some money of his own, but he did
not buy *Voortrekker*. She was sponsored by the Voortrekker
Trust, who said before the race that if she and Dalling did
well, they planned an assault on the other major ocean prizes,
their ultimate goal being the capture of the America's Cup
by a South African boat. Dalling was, in fact, one of the
handful of competitors in the race who did not own the boat
he sailed.

There was a great deal of comment and adverse criticism
about sponsorship in the 1968 race. Much of it was ill-informed, and born of a certain bigoted yachting conservatism
– a conservatism, it seems, found mostly amongst landbound
yachtsmen. There was even the suggestion that *The Observer*
or the Royal Western Y.C. should ban from the race all
sponsored boats.

The Royal Yachting Association at its annual general
meeting in March discussed the whole question of sponsorship and amateurism. But no conclusions were reached,
except that there was a general plea for restraint.

It was also hinted in the Press that the race was in grave
danger of being overshadowed by the competition to find
the wealthiest sponsors.

Williams suggested, when talking about the sponsorship
of *Lipton*, that with financing from commercial sources
there would be a stimulus to the development of gear and
racing-boats that would in time prove of benefit to the week-
end sailor in his cruiser. But he went on to say that, largely
because of sponsorship, the race would be two-thirds won
before ever the boats got to Plymouth.

The latter suggestion is not entirely valid. It is inevitable
that the biggest, fastest, best planned and organised boat in
the hands of even a moderate sailor will get to Newport first,
and therefore be declared the outright winner. But the

handicapping system provides for the smaller boat with the less financially resourceful owner. The satisfaction of the skippers of small boats like *Opus* and *Silvia II*, who must have known in Plymouth that they would never make the fastest crossing, must, in their own modest ways, have been quite as great as that of Williams. On corrected times both those boats made outstandingly good passages, making optimum use of the resources both of sailor and boat.

Other entrants in the race would have liked to have had some form of sponsorship. Sandy Munro could find none for *Ocean Highlander*, even though he was willing to change the name of his boat. Bevan looked for a sponsor, and in the end did very well without one. And I do not believe that the race will ever be overshadowed by the counter attraction – an open season for sponsor hunting.

The challenge of the race is still the challenge of the Atlantic. Man and boat against ocean. It is possible that the man with the greatest resources will be the outright winner. But there will always be boats like *Opus* and *Silvia II* and *Dogwatch* in the race, because it is obvious that their skippers are not merely interested in the silver salver that is *The Observer* Trophy.

By the beginning of 1968, the competitors' preparations were well under way. The organising machinery which had been ticking over quietly for the past four years, answering inquiries from prospective entrants and vetting possible boats, now moved into top gear as the Royal Western Y.C.'s Race Committee made its preparations.

An informal party was held in March where many of the prospective competitors were able to meet one another, and also some of the race officials. A guest at the party was Peter Dunning, the owner and manager of the Port o'Call Marina in Newport, Rhode Island, who when he returned to America was to be co-opted on to the Ida Lewis Race Committee as adviser and consultant.

The plans of Odling-Smee's committee for handling the

arrival of competitors in Plymouth were well in hand, and the Royal Western Y.C. was prepared to organise the start for a large number of boats (by March there were thirty-two entries), and by liaising early on with the organisers in Newport, Odling-Smee and Terence Shaw (the secretary of the R.W.Y.C.) hoped that few of the 1964 problems would recur.

The Race Committee in Newport met on 20 March, under the chairmanship of William Thomas, and comprised Thomas, Dr Fred Alofsin (the Mayor of Newport), William Muessel (the commanding officer of Castle Hill Coastguard Station), Dr Baldwin Sayer, and Dunning (who had just returned from London). Also present was Arthur Murphy (the secretary of the Ida Lewis Y.C.) a charming young Newport lawyer, whose help was to prove invaluable when I opened *The Observer* Press Office in Newport.

It was decided at the meeting that the privileges of the club would be extended to each participant during his stay in Newport, that there should be an office for the Royal Western Y.C. at the Ida Lewis Y.C., that the Press Office would be situated at the Port o'Call, and that a close liaison would be set up with the Yachting Committee of the Newport Chamber of Commerce. Radio frequencies were also to be made available for competitors to report their positions via Castle Hill Coastguard Station.

This time, Newport was very aware that there was to be a transatlantic race, and the town was well prepared.

Through the early months of the year the interest in the race increased. Yachting enthusiasts all over the world paid considerable attention to what promised to be a thrilling event. By May there were forty-three entries from ten different countries.

Then someone, it must surely have been in a newspaper article, suggested that the race would settle once and for all the controversy between multihulls and monohulls. And the arguments began in earnest.

Multihullers had been given a great boost in 1966 when Kelsall and Martin Minter-Kemp sailed the fast trimaran *Toria* to victory in the Round Britain Race. Until then, trimarans had been going through the unhappy teething period that catamarans had suffered a few years earlier. The victory of Kelsall's trimaran in a long offshore race was seen as the turning point, and trimaran devotees (who are no less emotional and illogical than any other class of yacht fancier) started to convince themselves that a three-hulled boat would win the race.

On the other hand, monohull men still believed that anything more extravagant than a single hull was, and always would be, basically dangerous. They were deeply saddened, but claimed to have been proven right, by the loss at sea in April-May of Arthur Piver.

Piver, as I have mentioned, had intended to enter both the 1960 and 1964 races in trimarans of his own design. He was a famous, and highly respected (even by those who did not agree with his views about multitudes) designer of yachts. He set ou, in a Dart-class trimaran on 17 March to sail from San Francisco to San Diego, a voyage of some six hundred miles, and expected to return to port by 22 April. Nothing was heard from him, and no ship reported sighting him. On 12 May the news filtered through that the San Francisco Coastguards had abandoned their 60,000-mile search for him, and that he was presumed lost at sea.

There were at that time twelve trimarans entered in the race, one of them – *Amistad* – a Dart-class boat. The multihull critics became more vociferous.

Suddenly the yachting world and many people outside it, were brooding on the danger of multihulls. The argument seemed to be that they were unstable and likely to capsize, or that they tended to break up – as if monohulls had not been known to sink, or cars to crash, or planes to fall out of the sky.

The critics of multihulls, like the critics of anything else, started to magnify the significance of isolated incidents. They quoted the loss of Piver and of Hedley Nicol, a brilliant Australian designer, who was drowned last year sailing in the Pacific.

The emotional arguments may appear to be strong. The statistical arguments are as follows. *Boating Statistics*, a U.S. Coastguard publication, shows that there were 1,318 deaths as the result of boating accidents in U.S. coastal waters during the year 1966. Multihull boats made no contribution at all to the fatal statistics. Obviously there are more conventional boats sailing in American coastal waters than there are multis; but the figures do not lie.

The best of boats, be they mono- or multihull, can capsize, or sink, or run aground when improperly handled. Surely the criterion for any vessel must be whether it is safe when properly handled by an experienced sailor.

Much of the controversy about multihulls is purely emotional, with the purists becoming outraged in much the same way as they were when Bermudan rig became fashionable. Designers of multihulls claim phenomenal speeds for their craft (speeds that are not always borne out) and exaggerate the good points of the boats, while the monohull brigade dwell only on the bad.

It was wrong to suggest before the race that it would be the great contest between multihulled and monohulled boats – nothing, especially yacht design, is ever so cut and dried. The Atlantic crossing from east to west is hardly a fair test for multis, especially when sailed only with a crew of one (as Bill Howell was graphically to explain to me after the race). It is said that it will prove very difficult for a multihull to win this race – the winds are unfavourable and a solo sailor has to spend too much time reining in the tri or cat to be able to sail at either his own or the boat's optimum efficiency. But then it must be argued that *Cheers* could quite easily have

Myth of Malham, *Noel Bevan's unsponsored ocean racer,
displays a classic cutter rig. Built in 1947 she
was the oldest yacht to sail in the Race.*

'The force of the wind and the manpower of the crew'
are allowed by the rules. So Stephen Pakenham urges
Rob Roy along with oars at the windless start.

*Mungo Campbell's trimaran Coila being sailed by
Eric Willis who was also responsible for the design.*

The heaviest yacht in the Race, Spirit of Cutty Sark.
*More than enough for most men, and a
handful even for the quiet giant Leslie Williams.*

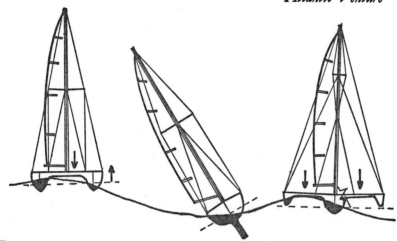

Fig 1

Sailing out of the page, as it were, with the wind blowing from right to left, are the catamaran, the monohull and the trimaran. As the wind increases, so the monohull heels over. The cats and the tris remain upright, presenting far more sail area to the wind. This gives them a greater speed potential. In a sudden gust of wind, the monohull heels further, but if the cat or tri has too much sail on her, then the strains are not eased by the safety value of 'heeling'. Terrific pressures develop on the rigging stays which hold the mast upright. Sometimes the mast will break. Or suddenly the whole boat will be pushed on to its side and capsize. Because of this danger, multihulls tend, especially when sailed singlehanded, to carry less sail area than a monohull.

won, and was very close on the heels of *Lipton* and *Voortrekker*.

In answer to that, and with the utmost respect to Newick who is an enormously successful designer, I would say that it is difficult to judge whether Tom Follett or *Cheers* came third. Follett would say 'both' – but then Follett would never see, admit or agree that he was by a dozen cables the best sailor in the race.

Golden Cockerel and *Gancia Girl* did as well as could be expected in the circumstances, and no worse than boats of comparable size and sail area. It is true that only five multi-hulls finished in the race, and that eight dropped out. But eight monohulls retired, and none of them was so revolutionary or untried as *Pen Duick IV* and *San Giorgio*.

In spite of the two multihull sinkings during the race, those who looked to the event as a testing ground for the two, or rather three, types of yacht one against another, must still be disappointed. Multihull design has not advanced nearly far enough for the boats to be put to the singlehanded test. When efficient self-steering gear has been fitted to big ocean-going multihulled yachts, and when it has been proven that a Tabarly can really handle so big a boat on his own in uniquely variable winds, then the battle might be rejoined in mid-Atlantic.

3
The boats and their trials

Of the thirty-five boats which set out from Plymouth, twenty-two were monohulls, four were catamarans and nine were trimarans. Some, of course, were older boats, not specially built for the race – and amongst them were proven ocean racers like *Myth of Malham* and *Gancia Girl*. There were newly built yachts of standard design – *Spirit of Cutty Sark* and *Gunthur III* – and some extremely unusual prototypes.

Four boats were, to my mind, unique, extraordinary and quite exceptionally fast – *Cheers*, *Pen Duick IV*, *Sir Thomas Lipton* and *Raph*.

The French seem to have mastered the technique of building with aluminium, which is at once strong and light – the strength–weight ratio is vital in a transatlantic race.

Mathematically, a yacht's speed potential rises as the length of its waterline increases. Competitors this year wanted bigger and bigger boats, and it is significant that both in 1960 and in 1964 the longest yacht won the race. Speed potential must, however, be coupled to the driving force, or engines, of the yacht. In the race, the only engines permitted by the rules are the sails and the efforts of the yachtsman – the limiting factor being the physical strength of the lone sailor. If he can handle five hundred square feet of sail at a time, then the sail plan must be designed around that factor.

The sails have continuously to drive a reluctant hull through a 'glutinous' sea. The lighter the hull, therefore, the easier it is for the sails to drive, and the faster the boat will move through the water. The winning combination, as has already been stressed, is the greatest possible length at the waterline, plus the largest manageable sail area, plus a hull of light displacement.

In their search for lightness, the French have turned to aluminium. Tabarly with *Pen Duick IV* and Gliksman with *Raph* built in Duralinox – which is extensively used in aircraft construction.

The three alternative forms of construction that Gliksman considered were 'bois moulée, l'alliage léger et le sandwich polyester'. Wood he felt to be weak on the superstructure and the sandwich method provided problems in France – 'The serious inconvenience is the novelty of the process.' So he decided on a light alloy.

Tabarly also turned to metal. His strong belief (amounting almost to an obsession with him) in the prerequisite 'lightness' caused him to abandon his well-tried and successful plywood construction. What is less understandable is his decision to build a multihull for the race.

The Frenchman was known to be a monohuller, and had written in 1964 that he did not think that multihulled boats were the right kind of craft for singlehanded sailing. There were perhaps two factors that governed his choice of a trimaran.

After the Fastnet Race in 1967, Tabarly went aboard *Toria* and showed considerable interest both in her form of construction and in the fact that a trimaran of comparatively short length and small sail area (*Toria* was then rigged as a sloop) could have won an offshore race in variable conditions. He asked many questions about the boat, sailed her, and went away, as Minter-Kemp put it, 'Clearly very impressed.'

It is also possible that Tabarly realised that he had made a mistake with *Pen Duick III* in building her too big for one man to handle – she had swept the ocean racing board in 1967, but with a crew usually of nine. Last spring he had been sailing her alone off the French coast, and when he tried to bring her into Lorient he simply could not get her round. The rumour persists that *Pen Duick III* had to be towed in.

Having decided on a trimaran, Tabarly typically took no half measures. She was to be sixty-five feet overall, ketch rigged and quite revolutionary. Her masts were to have a core of aluminium with an outer glassfibre skin, and the space in between was to be filled with Polystyrene foam. These wing masts were also designed to rotate, or rather revolve from side to side, with the sails. The tracks were completely semicircular allowing a gybe almost through 180°, and with the wing masts the strain on the sails and sheets was greatly reduced.

The development of the floats was intriguing. Nine different designs were tested, and it was established that the best floats were those which caused as little turbulence and wake as possible. Hence *Pen Duick*'s long riggers with fine ends and symmetrical section.

Not only was she the longest boat, but she also had the greatest beam – thirty-five feet. This was not simply to make her the biggest yacht in all directions: the metal stays joining the floats to the main hull were each nearly fourteen feet long, and were designed to give a float buoyancy of nearly five tons – the same as the total weight of the boat. In theory, therefore, if she capsized Tabarly would have been able to right *Pen Duick* again.

As soon as Tabarly arrived in Plymouth, he made it clear that he was not happy about sailing the boat after only inadequate testing. Her sea trials were skimpy, and most of his knowledge of her performance was gained during the all too short crossing from France to Devon.

'Certainly she is fast,' he reported, and he thought that at times her speed had been around twenty knots.

'She is not tried enough, and I have not had the time to bring her up in all weather conditions. I am not very confident.'

He added, 'Mind you, I was not very confident last time – and I won.'

According to Gliksman '*Raph* was not an especially simple boat' – though she seemed brilliantly designed for ease of handling by one man. Her measurements corresponded almost exactly with those of *Sir Thomas Lipton* – a foot longer overall, she was only nine inches longer than *Lipton* at the waterline, and her beam was a foot and a half greater than the English boat's. The draft, sail area and displacement of the two boats were similar, *Raph* being slightly the heavier. Their speeds must almost have been identical, and in the hands of much the more experienced sailor there is no doubt in my mind that, before she had to retire, *Raph* was in the lead off Newfoundland towards the end of the race.

Robert Clark, who designed *Sir Thomas Lipton* and also gave Williams much help and advice in the planning stages of the project and throughout the course of its development, started by studying the sail plan, as he maintained that the size of the boat would be determined by the amount of sail Williams could handle alone. The other considerations were that the boat had to be long in the water, yet cheap to construct, and above all simple.

For someone who is not really qualified to judge, it might seem impertinent to suggest that *Lipton's* concept was brilliant, her building superb and her success outstanding. But Clark, Williams and the boat all worked in close harmony, and with the building methods and help of Kelsall, they swept the best products of world famous yacht designers off the Atlantic. For Clark this was no new venture, nor an unexpected success – he has been designing winning boats for decades. But his pleasure at *Lipton's* success must have been especially keen, in that he was clearly delighted to be able to design a yacht that was not bound by constricting rules and ratings.

'Although the yacht was to be a racing yacht, the racing

38

was to be of a kind long forgotten – without handicap or rating. No thought of 'measured' sail area; no purpose in crowding headsails of extravagant overlap into a restricted foretriangle.'

Williams had decided that headroom was of no importance below deck – 'I just wanted to sleep and work there, not hold parties' – so Clark was able to make the freeboard only three feet.

'One is so used to the very high sided yachts which have been developed by the R.O.R.C. rule that it took some courage to settle for a freeboard amidships of only three feet.'

With no doghouse or large well to clutter the deck, the result is a long low hull and the general appearance of incredible beauty and swiftness through the water.

Lipton was built at Kelsall's yard in Sandwich by the 'sandwich' foam-G.R.P. (glass reinforced plastic) construction method. The sandwich in the method refers to the mode of construction and not the place in Kent. Kelsall pioneered the form of construction in this country, and proved its strength, lightness and seaworthiness when he sailed *Toria* to victory in the Round Britain Race.

The principle is a simple one whereby a strip mould (shaped to order for each yacht) of upright frames and longitudinal stringers is covered with sheets of P.V.C. foam. An outer skin of glassfibre is laid over the mould and finally the inner skin is laid to complete the 'sandwich'. The method is relatively cheap and speedy, and provides a tough, light hull.

Lipton's deck was made of glassfibre-covered plywood, and added very little weight to the boat. Clark described Lipton as 'an extreme fin and skeg type, a concept which at first filled me with horror'. He feared the erratic steering of such boats, though Lipton seems to have had few steering problems.

Like many other of the yachts in the race, she was fitted with Hasler self-steering gear – the wind vane gears invented, pioneered and developed by Blondie Hasler and manufactured exclusively by M. S. Gibb Ltd. More than half the starters chose Hasler gears for their boats, and of the first ten to finish, seven had been steered across the Atlantic by Hasler. Except for *Cheers*, which did not have self-steering gear, various types of Hasler gear were fitted to the first six boats home.

When the construction of *Lipton* was finished, Williams sailed her to Cowes, from where he planned to start his qualifying trip and the yacht's pre-race tuning. In the entrance to Cowes harbour, she was in collision with a hovercraft piloted by an officer of the Royal Iranian Navy.

'In fact we were rammed,' said Williams, 'it was entirely the fault of the hovercraft.'

Lipton's lee rigging was damaged, but the hull was not holed. 'The hovercraft sustained considerable damage,' joked Williams.

Repairs to the rigging kept the boat out of the water for a couple of days, but then he took her from Cowes to the French coast, across to the Lizard and into Falmouth in less than three days, at an average speed of around seven knots. *Lipton* was fast, and Williams had sounded his warning to the other competitors.

It was a warning clearly heard by Dalling, who was already in Plymouth making *Voortrekker* ready for the race. Dalling had sailed the boat from Cape Town to England, with a crew of two, in very good time, averaging a hundred and sixty miles a day over the seven thousand mile voyage.

The logic behind the design of the South African yacht was (again) that the governing factor would be sail area. Her designer, van de Stadt, considered that one thousand square feet of sail area was the maximum that one man could handle – a sail area probably less than some of the other competitors'. So to have a comparable waterline length and thus effective operational speed, he designed a hull that could be driven by

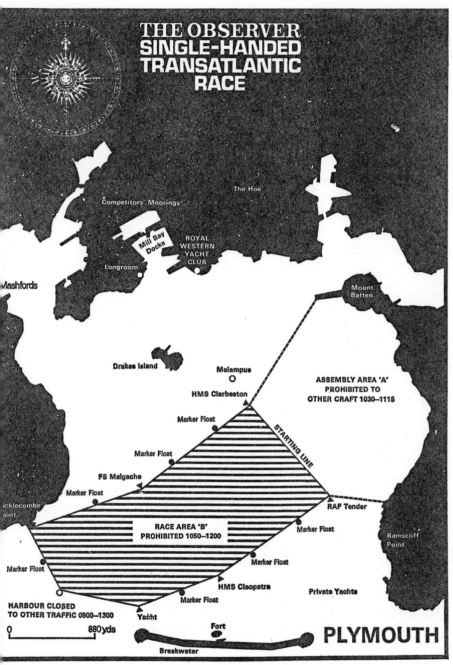

Fig. II The Plan of the Starting Area.

a relatively small sail area. The result was a ruthless end product with a displacement of only six and a half tons. The hull was constructed of triple laminated mahogany, and the weight concept was carried through not only in the hull but with the fittings as well.

In the end, the saving of weight may have proved to be disadvantageous – three of the lightweight winches were to break during the crossing to Newport, and all Dalling's winches gave endless trouble in the heavy conditions.

Voortrekker also was a fin and skeg boat – one of the rare occasions when van de Stadt has given one of his designs a skeg-rudder – and Dalling testified to her directional stability and her ability to 'surf' once the seas rose to a reasonable height.

'Even under her full two thousand square feet of running sail she never gave any indication that she would broach.'

By no stretch of the imagination could she be called a handsome yacht. She seemed dumpy and workmanlike, certainly by comparison with *Lipton* when she was moored next to the English boat in Newport after the race.

Dick Newick may be less well known to the general public than Clark and van de Stadt, but he is certainly a designer no less accomplished. His successes are considerable though apparently less widely heralded than others'. He is a multihull man, and to him was entrusted by Jim Morris – the co-ordinator of the *Cheers* project group – the task of devising a yacht in which Follett could win the race.

There was but one purpose for *Cheers*, and that was her participation in the last race with the objective of winning. The three men, and their helpers, completely dedicated themselves to that endeavour, never compromising in her design or construction and ever considering her seaworthiness.

The idea for *Cheers* was entirely Newick's, and she was created on a blank sheet of paper with the aim that she would be unique in theory and design and exceptionally fast. The

design was based on the old Pacific islands outrigger canoe –
though *Cheers'* main hull was intended always to be kept to
windward. The equal hulls were similar at both ends (she
had no conventional prow or stern) and the rig could be re-
versed to allow her to be sailed in either direction. That
manoeuvre took longer than normal tacking, though once
under way, her speed more than made up for the time lost
in setting a new course. She was constructed of triple
mahogany planking and her hulls were covered with poly-
propylene cloth and epoxy resin. The project was unsponsored by any organisation or commercial enterprise.

Early this year during trials in the Caribbean, *Cheers* capsized, 'It looked as though we would have to scrub this
particular venture,' said Morris. The Royal Western Race
Committee had already indicated that the yacht was almost
certainly 'not acceptable'.

Newick informed the Yacht Club of what had happened,
and remedied the fault by adding a 'blister' to the windward
hull, so that should *Cheers* again be caught a-back, the added
buoyancy would hold the masts well out of the water while
the sails were taken off and the yacht righted. But the Race
Committee still refused to accept *Cheers'* entry, though they
gave a provisional acceptance.

Follett then sailed the boat from the Virgin Islands to
Gosport (a distance of four thousand two hundred miles),
in just over twenty-eight days and her Atlantic-worthiness
being proven, the Race Committee officially accepted the
entry.

Both Odling-Smee and Shaw have emphasised that the
Cheers team never pressed their case dramatically nor did they
take offence when they were persistently refused. They seem
to have been remarkably patient, realising that *Cheers* was a
revolutionary boat whose ability and seaworthiness had to
be proved. They asked no favours, and even after they had
arrived in England, they said that if the committee still

43

refused to accept their boat they would abide by the decision – a welcome change from those competitors who said that if they were turned down, they would sail anyway, and 'damn the Royal Western'.

There were a welcome number of production line glass-fibre boats in the race. As resistance to G.R.P. has ebbed, so builders have been able to make larger and faster yachts.

Spirit of Cutty Sark was one of the Gallant 53 Class, and her speed and seaworthiness can be judged from her position in the race – though it must be added that she is really too heavy a boat for singlehanded sailing, and I doubt if anyone less capable than Leslie Williams could have handled her. At the other end of the scale in size were the glassfibre sisterships *Fione* and *Goodwin II*, displacing only one and a half tons. *Aye-Aye* was also built with G.R.P. – she being a Hanseat Class yacht.

Ocean Highlander was an Ocean Ranger 45 Class catamaran, *Maguelonne* an Elizabethan 35 boat and *Ambrima* a Giraglia. The great advantage of a production boat is that the yachtsman knows beforehand what the yacht will look like. William Wallin ordered his Invicta Class sloop *Wileca* by post, and only saw her for the first time a few weeks before the start when he arrived in England to collect her.

Other than *Jester*, the old ladies were *Zeevalk* and *Myth of Malham*, which was built in 1947 and was at one time a member of John Illingworth's famous Malham fleet. In 1949 van de Stadt designed *Zeevalk*, the first of his yachts to have a separate rudder track – *Zeevalk* was the fourth of the Dutchman's designs in the race, the others being *Voortrekker*, *Wileca* and *Spirit of Cutty Sark*. *Opus*, another Clark-designed boat, was registered at Lloyd's as a 1965 yacht, though she was really of a 1957 vintage, having taken eight years to build.

Most of the competitors had radio receivers on their yachts and some had high-powered transmitters as well. Geoffrey

Williams needed his to communicate each day with his sponsors, and others had daily or weekly radio rendezvous with various newspapers. Radio contact was the only reliable means of ascertaining the whereabouts of a boat, as there was no guarantee that competitors would be sighted by passing ships or, if they were, that their positions would be reported. All the yachtsmen were required to fly international recognition pennants – and yet there were no reported sightings either of *Cheers* or of *Jester* throughout their voyages.

For radio communication the lone sailors and the Press and sponsors were dependent on the G.P.O., and the work of the Office both in London and especially at Baldock Radio Station was tremendous. At Baldock they listened in on the appropriate frequencies almost day and night, and the quality of the transmissions was always better than the conditions allowed.

During one of our conversations, Minter-Kemp remarked that the G.P.O. telephone system seemed to be more efficient if you were in mid-Atlantic than if you were trying to call Flaxman from Kensington.

Towards the end of May, all the boats, new and old, glass-fibre, wood and aluminium, were being tuned for the race. In no time at all, Plymouth was invaded by the motley and colourful armada.

4
The competitors assemble

Millbay Dock was never designed to be a yacht marina, yet the deep, dirty basin had to serve as one before the race. It was essential that a central mooring place be provided for the competitors' boats, where they could be provisioned and worked on, and also viewed by the crowds of spectators who came to watch the strange collection of rather untidy, mostly bearded seamen prepare for the epic voyage.

In the weeks prior to the race, a great deal of criticism was levelled at Plymouth City for the conditions in the dock – though none, it must be added, was made by any of the competitors. Local pressmen and non-participant yachtsmen were the most vociferous in their outcries, their main point being that a sea city, now famous as the starting place for many off-shore and ocean yacht races, should have a proper marina with adequate facilities.

The Docks Board only accepted the yachts because there was nowhere else for them to go, and the Royal Western Y.C. realised that it would be necessary for competitors to assemble at some central place so that they would be easily accessible to the Race Committee offices and the Press office, and where their boats could be satisfactorily scrutineered.

Obviously the conditions in Millbay Dock were unsatisfactory. Some boats were moored two or three abreast, the water was filthy with oil slicks, flotsam and all manner of debris. At the western end of the dock, dust from the coal tip covered the quay right to the water's edge and in the slightest breeze was blown on to the decks of the boats. Competitors were constantly having to sweep their yachts clean, and one

Frenchman was even seen dusting the quayside with a large broom.

In spite of the inadequacies both of the site and the amenities (the Royal Western Y.C. provided huts with washing and shower facilities and lavatories on the quayside) Shaw explained that without the co-operation of the Docks Board, there could have been no race at all – for there simply was no other place with deep water moorings where so many yachts could have been assembled.

Situated at Millbay Dock were the Race Committee office, *The Observer* Press office and the Yacht Reception office (manned by employees of the Plymouth Gin Company, whose work on behalf of the competitors was tireless and whose praises, strangely, have hardly been sung).

Ruth Shaw helped her husband in the Race Committee office, and hers was perhaps one of the most tiring vigils. The organisation of the Yacht Club was meticulous, both in arranging the start and in assisting those competitors who shortly had to return to Plymouth for various reasons. This work was done so quietly and unobtrusively that there was always the danger of the club having appeared to do less than in fact they did. The competitors all testified to the kindness of the club and of the Shaws in particular.

Early in the year, *The Observer* had appointed Mathers Public Relations to handle Press liaison and to run the Press office. Guy Pearse of Mathers invaded Plymouth with his minions some eight days before the race started, and they manned the Press office almost day and night. At the same time, Malcolm Turner set up an operations room at Mathers in London from where, during the course of the race, all information about competitors emanated. Pearse was given considerable assistance in Plymouth by the staff of Westward Television, who took a great interest in what quickly became very much a West Country event.

To the dusty, but none the less convivial, atmosphere of

47

Millbay Dock then, all the competitors sailed. An early arrival was Dalling in *Voortrekker*.

Dalling had interrupted his studies at Natal University to take part in the race. He is reading a postgraduate course in philosophy and theology, having already taken a degree in agriculture. He was at one time in the Police Force in Hong Kong – where he took the trouble to master Cantonese. It was in Hong Kong that he learnt his sailing, and he returned to South Africa sailing his boat *Carina* singlehanded across the Indian Ocean, through the Mozambique Channel and down the east coast of Africa to the Cape.

In Plymouth, Dalling talked quietly and deliberately, treading delicately over the thin ice of politics and reserving all his enthusiasm and emotion for sailing.

'I don't think anyone should take on a singlehanded voyage unless he has a clean-cut purpose and plenty of reading and thinking to do.

'But after being alone you are in a state of euphoria – you love the human race – and then immigration people in hobnailed boots tear everything out of your boat and you end up as twisted as you started. You've wasted all that time dreaming of blondes and beer.'

Voortrekker carried a very powerful radio receiver, and Dalling said that he would listen to anything except pop.

'Something with emotional richness about it, like the Bruch violin concerto, is best for the sea. But not Wagner – Wagner is like eating something salty with salt.'

For reading matter, he decided to take poetry, a biography of T. E. Lawrence and something light '. . . probably a cheap, sleazy American novel'.

As for the other competitors, Dalling feared Tabarly most. 'But don't let there be any misunderstanding, I bloody well intend to win.

'Then, when it's all over, I shall probably buy myself a cattle ranch and lead a quiet life. The whole thing's been a

Bill Howell in Golden Cockerel. *She turned turtle last year, but was repaired and a taller mast added.*

Bertil Enbom sails Fione *through a blustering sea.
He reached Newport after 40 days at sea, a remarkable
passage for a 19-foot boat.*

Brian Cooke made a fast crossing in Opus. *With a cruising rig, cruising sails, and without sponsorship his was the unsung feat of the Race. The dust cover dramatically shows Cooke working the boat.*

*Edith Baumann, the only woman in the Race, seen
aboard her weird* Koala III, *complete with Turkish slipper
bows, anti-whale bars and, of course, her dog* Schatz.

hell of a shock to my system – I normally live in a very quiet academic atmosphere.'

The three Swedes who entered the race were Ake Mattsson, Bertil Enbom and William Wallin. They moored their yachts *Goodwin II*, *Fione* and *Wileca* together at the eastern end of Millbay Dock. The boats were all sloops, the biggest being the British-built *Wileca* at twenty-seven feet. *Goodwin II* and *Fione* were sisterships and at nineteen and a bit feet were the smallest in the race.

Mattsson was the most experienced of the trio, and he and Enbom – a Lieutenant in the Swedish Army – were both to do remarkably well in their diminutive boats.

Myth of Malham looked as graceful as ever at her mooring, and Bevan seemed justly proud of the unique detection equipment and self-steering gear he had himself designed. The forty-nine year old electronics engineer, who comes from Frimley, planned to use special correction tables to enable him to calculate his true position even in the worst conditions.

Pakenham exchanged his pulpit for the doghouse of *Rob Roy*, and the vicar's parishioners at Apuldram and Donnington, near Chichester, awaited anxiously the outcome of their pastor's venture.

'Seamanship is the best training for self-reliance,' said Pakenham, who was at one time in the Fleet Air Arm. He also took a degree in economics and theology at Cambridge, and with his dual seaman–clergyman personality, he was objectively interested in the human as well as the technical aspects of the race.

'I am afraid that the younger entrants may be too intense, expect too much of themselves and have nothing to fall back on if they find they're not doing well.'

He had introduced the idea of his proposed participation in the race somewhat indirectly when he wrote to his parishioners. 'Dear Friends, You will probably know that I plan to be away for three months this summer. . . . The conviction that I should take part in *The Observer*'s race came to me towards the end of a four-day retreat in January 1965. . . . I was duly surprised. . . . However, when guidance is granted to us, we must follow it, blindly if necessary. . . . It would be quite wrong to enter the race purely as a glorified retreat! If it is right to speak of upholding the honour of the Church, then I must enter wholeheartedly into the spirit of the race.'

Before the start, Pakenham seemed relaxed and had no anxieties about the other competitors. He was keen to win the Handicap Prize, and his wife was completely behind him. 'As long as I have regular news, I shan't worry. He is so self-reliant and cool in an emergency.'

Another expert seaman with the will to do outstandingly well was Eric Willis, a yacht deliverer who conceived the idea of *Coila*'s design. A firm believer in the ability of a multi-hull to win the race, he devised a quadrant with which to steer the boat rather than the more conventional tiller or wheel.

By far the most handsome yacht in Millbay Dock was *Spirit of Cutty Sark*. Leslie Williams joined the Royal Navy in 1948 and transferred to the Fleet Air Arm during his apprenticeship. Having trained as a Radio Electrical Artificer, he was appointed Radio Officer at R.N.A.S. Culdrose in 1966.

His sailing experience was gained mostly in Naval yachts, and he skippered and navigated the Naval Sail Training Yacht *Merlin* in the 1964 Tall Ships Race. So the Atlantic was not a totally unknown expanse to him.

The Lieutenant had represented the R.N.S.A. in many off-shore racing events, and served as Watch Officer on *Sir Winston Churchill* during her maiden cruise. He had also competed in the last Round Britain Race.

Williams is a quiet giant of a man, and with his considerable experience seemed a very likely winner. I thought that the race would probably be fought out between him and Dalling, though the odds seemed to be on the South African. At nearly sixteen tons Williams's sloop was by far the heaviest boat in the race, and, with only one mast, *Cutty Sark*'s sails appeared huge for one man to handle alone. Moreover, Williams's recently dislocated right arm was still in a sling, and was obviously giving him more trouble than he would admit.

The only Australian in the race was Bill Howell, a cheerful dental surgeon who has a practice in London. He, like Tabarly, had sailed in the 1964 race, and had also sailed solo from Plymouth to Panama and across the Pacific from Tahiti to British Columbia – when he earned the colourful sobriquet 'Tahiti Bill'. He seemed confident that he could win in his large catamaran *Golden Cockerel*, in spite of her size and speed and her having capsized last year.

The banking world had been campaigning for some time before the race started to prove that bank managers were approachable, charming, human people. Cooke, the sub-manager of Westminster Bank's Poole branch, also showed that some of them at least were extremely athletic and daring. *Opus* was lent to him for the race by her owner R. F. Austin.

Cooke was overmodest about his chances. 'The whole thing's recreation when all's said and done. Isn't it really? But I want it to be worthwhile.'

The Westminster Bank in Poole has a distinctly nautical atmosphere about it – yachts undulating across the curtains in the manager's office, marine watercolours on the walls, and a chart of Poole harbour in the waiting-room. Cooke was an inaugural member of the Bank's sailing club, and the club backed him in his decision to enter the race. 'I put the idea to the flag officers and they were agreeable as they knew I'd done a bit of sailing.'

He had in fact been sailing since he was a boy on the Norfolk Broads, and then throughout his service in the Merchant

Navy (he left with a first mate's ticket) and later obtained his yacht master's ticket.

'An ocean passage is something I've always had a sneaking wish to do, but I only dreamt of doing it when I retired, like most people who have loved the sea.'

The Bank gave him two months' paid leave of absence and its official blessing, and Austin offered to lend Cooke the boat he had helped build in a back garden in Stanmore. Thus a big bank went to sea.

Pyle's home-made ketch *Atlantis III* was a bit more do-it-yourself than *Opus*, and was not only to be sailed single-handed but was also built without help – '. . . though friends gave me a hand with the keel – I'm not much good at working in metal'. Pyle, who has been sailing since he was seven, built the boat at Calshot Activity Centre on the Solent where he was sailing and climbing instructor.

Amistad had already won the 1967 New York–Bermuda Multihull Race when Bernard Rodriquez decided to enter this race. His Dart-class trimaran *Amistad* was designed by Piver, and Rodriquez sailed her to Plymouth through one of the worst storms he had ever encountered.

The twenty-five-foot yacht encountered gale force winds almost all the way from New York, and the voyage took forty-two days.

'I thought there were supposed to be westerlies in the Atlantic. . . . The night of the storm the wind was the fiercest I have ever seen it. I peeked out the window only once.'

His steering vane was smashed by the pounding waves, but Rodriquez, who is a boatbuilder, was able to rig a new vane from scraps of plywood. Moored at Millbay between *Jester* and *Cheers*, the trimaran looked fragile and thoroughly scaring.

Edith Baumann, an attractive, extrovert twenty-six year old from Aachen, caused a sensation in the yachting world when she entered the race. Until she went to La Rochelle in the summer of 1967 she had never sailed in her life. While there on holiday, she met a retired French naval commandant, Waquet, and persuaded him to build her a trimaran and to undertake her sail training. She explained to him that she was feeling fed up, because she had been frustrated in her desire to be a helicopter pilot and explore the Amazon Basin.

Waquet designed everything for her safety: an anti-whale bar ('Poor whale,' said Edith) stretched between the prows of *Koala III*'s riggers and main hull, provided the first line of defence against mammalian interference. The bows of the main hull were packed with polyurethane foam, and the struts which joined the floats to the hull were reinforced with steel bars. These and the wire stays were designed to ensure that the floats could not be parted from the main hull and also to act as a bumper against flotsam. The Commandant fitted the yacht with hydrobrakes, in case she tried to 'run away' with Edith downwind. *Koala III* was designed to be both stable and unsinkable.

Edith's main worry was that she might put on weight during the race. 'My mother said she would die if I went ahead with this, but now she sees pictures of me looking well and fat, she feels better about it.' And for company, she had decided to take with her an endearing dog of uncertain origins. She acquired Schatz from a dogs' home in February, and before ever she thought of victualling her boat for the voyage had provisioned for her mongrel with a large case of dog food.

She was blissfully ignorant of what Waquet scornfully called 'the poetry of sailing', preferring to call her port float her left. But she was tumultuously welcomed at the Royal Western Y.C. when she sailed dramatically into Plymouth having completed her five hundred mile solo voyage. The officers of the club even allowed her dog into the bar – a privilege never before accorded to a canine. She became extremely popular with all the competitors before the start, and not simply because she was indisputably the prettiest entrant. 'She is fun and she is brave and I hope she wins,' said one of the French journalists.

Waquet, who has been somewhat unkindly called 'Baumann's Svengali', took his duties as tutor and mentor very seriously. An ex-pilot and ex-submariner, he was confident that he alone had solved the problems of trimaran design. 'The centre of buoyancy of the floats must be at forty-five degrees to the point of force on the mast.'

He also sailed a trimaran of his own design, though one rather smaller and altogether less apparently seaworthy than Edith's. His contempt for the yachting world knew no bounds, and he said with great conviction 'I have done more for yacht design in two years than your R.O.R.C. has done in fifty.' As it turned out, he may have been right at that – though not quite in the sense he intended.

'Edith knows that she will only get the green light on the thirty-first of May when I am satisfied that she is ready to go,' insisted Waquet.

The thirty-four foot trimaran *White Ghost* looked a rather ugly yacht moored beside much sleeker rivals. Michael Pulsford built her himself, and while she might not have been a visually pleasing boat, he proved her sailing qualities during a tough qualifying voyage to Norway and back. Pulsford, who comes from Scarborough, had a not untypical Yorkshireman's will to win.

The most famous Atlantic veteran among the starters in the race was not a man, but a boat. *Jester* was bought in 1965 by Michael Richey, who now had her moored at the far end of Millbay Dock.

Richey said that he would be sailing not so much against the other competitors as against *Jester*'s previous best time for the east–west crossing.

'I am basically non-competitive, but I want to do well in terms of what she has done before. If I can match Hasler's passage, it will be an achievement. *Jester*'s design is the result of Hasler's innate seamanship – he is certainly one of the greatest seamen in the world – and she is a boat to make the voyage in. Ordinary sailing bores me stiff.'

Colin Forbes chose a multihull for the race. The thirty-three foot sloop-rigged trimaran *Startled Faun* was built five years ago, and Forbes thought that she was at the same time sturdy and fast enough to make a speedy crossing. He also intended to record his passage on colour film – he is a film maker – in the hope that the adventure might be shown on television.

Dogwatch was another British entry, bought for the race by Nigel Burgess – a second officer in the Royal Fleet Auxiliary Service. He only returned to this country just in time for the start, having been serving on a ship in the Far East as navigation officer.

One of the youngest competitors was Robert Wingate, whose forty-foot sloop *Zeevalk* was built nearly twenty years ago. Before the race, he thought that the yacht had a leak, a fear borne out by subsequent events. In spite of his age, Wingate, who is a yacht deliverer with considerable knowledge of boats and sailing, hoped to do well, although, he said, the handicap system was not entirely fair on elderly ladies of the ocean like *Zeevalk*.

Perhaps the most experienced of the German trio was Egon Hienemann, who brought his thirty-one foot sloop *Aye-Aye* across for the race. His main worry was the self-steering gear, which he felt might not prove adequate for the crossing.

After the withdrawal of John Ridgway, the Army's sole representative was Martin Minter-Kemp – a Captain in the Royal Welch Fusiliers. The Regiment fosters athletic adventures, especially if they are less commonplace than cricket and rugby, and Minter-Kemp was given time off to compete in the race as part of his 'adventure training'.

He was lent *Gancia Girl* by her owner Tim Powell, though Minter-Kemp was of course no stranger to the trimaran.

The dark horse of the Atlantic was thought to be Sandy Munro, a tough, outspoken Scot from Inverness. He did not moor his forty-five foot catamaran *Ocean Highlander* in Millbay along with most of the other competitors, but kept her up at Devonport, next to Pakenham and *Rob Roy*.

Until 1967 Munro had been a pilot in the Fleet Air Arm, and on leaving the Service raised the £15,000 to buy *Ocean Highlander*. He said that he had plans to charter the yacht after he had won the race, though if he could find a sponsor, he would not have minded sailing solo round the world.

The three Frenchmen, Andre Foezon, Bertrand de Castelbajac and Lionel Paillard all sailed sloops. Foezon was the most experienced, and said that he needed very little sleep. He hoped to win the handicap prize in his fast, though comparatively small yacht, *Silvia II*.

Switzerland has never been renowned as a sailing country, but Guy Piazzini planned to change that by winning in *Gunthur III*. She seemed a sturdy boat, of the well-known Rorqual Class; her main drawback was that she was not designed for solo sailing.

The third German, though the first to arrive in Plymouth, was Claus Hehner in his thirty-seven foot sloop *Mex*.

If *Cheers* was controversial, Follet definitely was not. He had arrived in Plymouth in good time, and after his boat had been accepted by the Race Committee, spent his days visiting the other competitors on their boats. Not for him the final and frantic last minute preparations for the race – his challenge was brought to racing pitch and was ready to go days before the start. While Newick and Morris thought that *Cheers* was going to win, and they said so, Follett would not be drawn on the subject. All he would say was that he hoped to make a speedy crossing within the time he had set himself of twenty-one days. He was clearly more interested in proving his boat than winning a prize.

Follett must have been one of the most popular competitors, both with the other yachtsmen and the Press and with everyone who had anything to do with the race. He was patient in explaining how his strange craft worked, and would take spectators on board at all times to show them *Cheers'* simplicity. He talked about the race and the sea and almost anything that was mentioned. But for biographical details I had to go to Morris.

'Tom is a very knowledgeable man of the sea with tremendous experience, yet is so damn unassuming. He has a wealth of knowledge on so many subjects, from electrical engineering to gourmet cooking, flamenco guitar playing to languages.' Follett was a fighter pilot in the Second World War, and afterwards served as a radioman in the Merchant Marine. He has lived in many places in Europe and America, following a career in electronics engineering, and in the past few years had delivered a wide variety of yachts in both the Atlantic and the Pacific. He endeared himself to everybody by his humility, charm and droll wit. But it was not until after the race, when Howell sailed *Golden Cockerel* into Newport, that the right descriptive phrase was coined. The Australian said of Follett 'He was truly the superman of the race.'

Yaksha was very, very ugly. She was designed by Joan de Kat, and built quite quickly for the race. The astonishing thing was that even after the short and calm crossing from France, when the heavy metal float stays on the yacht were twisted and snapped by the force of the waves, de Kat did not realise that his trimaran was really totally unsuitable for the Atlantic crossing.

'She is not very strong,' he admitted, 'and in rough weather the floats smack the waves and send shocks through the boat.' But, being a Frenchman ('I am also an artist') he shrugged off misadventure.

'I will win,' he said, and grinned hugely.

Yves Terlain and Marco Cuiklinski were two of the luckier

Frenchmen who got their boats *Maguelonne* and *Ambrima* to Plymouth before the strike in France paralysed the country.

Tabarly was less fortunate, and after all the trials of getting *Pen Duick IV* into the water, the troubles in France must have seemed the final blow.

Political considerations aside, the riots across the Channel caused grave anxiety in Plymouth. While the students of the Sorbonne were chanting the names of Ché Guevara and Dany the Red in the streets of Paris, the organisers of the race in Plymouth, and the other competitors, began to wonder when, if at all, Tabarly and Gliksman would arrive in Devon.

Monday 27 May was to have been the qualifying date for competitors, and, under Race Rule 11, all yachtsmen should have completed the five hundred mile solo voyage and presented themselves and their boats at the Royal Western Y.C. by then.

Because of the unrest in France, Shaw sent out a memo explaining that if the latecomers had a valid excuse, Rule 11 would not be enforced.

This news must have brought some joy to Alain Gliksman, for the work on *Raph* had been held up by the strike. He said that he would try to move his boat to Plymouth on the Tuesday, and he in fact arrived late on Tuesday evening. The rumours about Tabarly continued to abound, and everyone said that it would be a great pity if he could not get to England in time to defend his title.

He eventually arrived, exhausted and with injured ribs, bruised by a runaway boom, with a far from ready yacht, early on the Thursday morning. Tabarly and his helpers worked on *Pen Duick IV* with frantic energy to try and bring her to racing pitch before Saturday's start. They were watched with fascinated interest by hundreds of spectators and wellwishers, who gazed with awe on the silvery machine that was favourite to win the race.

Gliksman, the modest though debonair editor of France's

premier yachting magazine, worked no less hard on *Raph*. After Tabarly's, his was the second longest boat in Millbay, and she looked beautifully designed and easy to handle. He said that he hoped to do well, though the yacht had not undergone adequate sea trials. Of the other competitors he feared Tabarly most – even in his barely finished trimaran.

'Untried boat? Ha! Eric could win the race at twenty knots on a piece of bread.' And he genuinely admired Tabarly and respected his ability as a sailor.

The other foreign entrant who was able to avail himself of the amendment to Rule 11 was the Italian Alex Carozzo. Carozzo, appropriately, comes from Genoa, and he designed and built the ballroom-sized catamaran *San Giorgio* himself.

The Italian Ministry of Shipping had offered to arrange the transport of *San Giorgio* to England. But the general election in Italy had caused a general attack of post-electoral inertia – all Carozzo's efforts to obtain the promised help were a dismal failure.

In despair he asked about in Genoa and discovered that the Currie Line (a British shipping company), whose boat the *Gisela Russ* was bound for London, would be willing to ship the catamaran as deck cargo. The *Gisela Russ* was to sail on 21 May from Palermo. *San Giorgio* was in Venice, whence Carozzo hastened, and he set sail for Sicily with a crew of two.

To start with they made good headway, but lost the wind off Brindisi and were becalmed. At this point the Italian Navy were ordered to lend a hand and *San Giorgio* was towed first to Messina and then on to Palermo. The yacht was finally stowed as deck cargo on the ship.

The *Gisela Russ* was not due in London until 29 May, which hardly left Carozzo time to get his boat to Plymouth for the start on 1 June. It seemed as though the Italian's efforts were to be frustrated yet again. But after some judicious telephoning from London, the Currie Line nobly agreed to put in at Plymouth and discharge the nautical cargo there.

On Tuesday 28 May the cargo ship arrived in Plymouth and *San Giorgio* was taken off with the aid of the Navy – the cranes in the civilian dockyard were not big enough to handle the unloading.

Carozzo still had to step the masts and rig the catamaran, and his work was not completed until after the other competitors had set out for America on the Saturday.

Clearly the Race Committee were not prepared to penalise the two Frenchmen and the Italian, all of whom arrived late through no fault of their own. Geoffrey Williams's was a different case altogether.

When it had seemed likely that some of the competitors would moor their boats at local yards rather than in Millbay Dock, Shaw had sent out an explanatory memo (on 22 March) to all entrants reminding them that they had to present their yachts at Millbay for scrutineering by Monday 27 May. There can be no question but that Williams received the notification after he had entered on 26 April, because not only did Shaw post him the memo, but he also telephoned him later in Falmouth where *Lipton* was undergoing some electrical repairs to tell him of the ruling.

Williams then brought his yacht to Plymouth, and slipped her at Mashford's Yard. He said that he feared *Lipton* might have a leak, and she had to be put on the slip to investigate. It has not been confirmed whether or not the boat did have a leak, and as some of the other competitors pointed out, while she was on the slip the underside of her hull could be defouled and polished.

A dirty and fouled bottom can make much more difference to the speed of a yacht through the water than is often imagined, and while Lipton was high and dry the other yachts were moored in the filthy water of Millbay.

The scrutineering took place in the dock on the Tuesday and Wednesday, and Williams did not present his boat to the committee, nor was there any word of explanation from him.

Odling-Smee nevertheless sent a member of the scrutineering committee across to examine *Lipton*'s seaworthiness, and he returned with no mention of a leak. When the committee decided to meet to discuss the possible imposition of a penalty on *Lipton*, Odling-Smee asked Williams to give in writing his reasons for being neither in Plymouth by the Monday evening deadline nor in Millbay for the scrutineering, and Odling-Smee told me that in his report Williams made no mention of a leak.

So the Race Committee met, and imposed a twelve hour penalty on *Sir Thomas Lipton*. Odling-Smee said:

'Everybody else obeyed our instructions except Geoffrey Williams and so we had to impose a penalty though we were rather loath to do it. We imposed what was really the minimum penalty.'

Before this meeting of the committee, there was, however, the competitors' briefing in the Royal Western Y.C. All the thirty-five starters, including Geoffrey Williams, were there, and details of the start arrangements and the facilities for towing out of Millbay by the Royal Navy were given to the competitors by the Queen's Harbourmaster, Captain Keith Welch. The R.A.F. provided a weather synopsis for the coming forty-eight hours, and Pearse discussed the arrangements for broadcasting reports from the Meteorological Office and the G.P.O. communications transmissions.

Shaw then went through the sailing instructions, and drew attention to an important omission in the section headed 'Course'. The paragraph read 'Cross the Starting Line from North to South. Leave Melampus Buoy to Starboard and Distance Mark to Port. Thence to Newport, passing South of Nantucket.' Some of the competitors' instructions had been amended to read 'passing South of Nantucket Light'. And Shaw asked all the yachtsmen to add 'Light' if it had not already been inserted.

It was the impression of the officials at the briefing that all thirty-five competitors were present when Shaw made his request; which, in the light of subsequent events, is an important point.

After the decision to penalise *Lipton* was made public, she dropped from being second favourite to sixth and her odds lengthened from six to one to tens. Shortly before the start, Ladbrokes had opened a book on the race – the first time ever that an internationally famous firm of bookmakers had offered odds on an offshore yacht race. Bets were accepted for the outright winner and not for handicap placings.

When the boats were scrutineered in Millbay Dock, all the monohulls were given a 'handicap factor' by which handicap placings could be worked out at the finish and corrected times established. The corrected time was worked out by multiplying each yacht's elapsed time by the factor.

	Yacht	Factor
42	Raph	0·8871
2	Voortrekker	0·8544
9	Spirit of Cutty Sark	0·8534
23	Zeevalk	0·7989
6	Myth of Malham	0·7890
38	Mex	0·7721
37	Silvia II	0·7378
39	Maxine	0·7281
45	Maguelonne	0·7233
7	Rob Roy	0·7218
47	Ambrima	0·7201
14	Opus	0·7184
31	Dogwatch	0·7080
16	Atlantis III	0·6950
40	La Delirante	0·6930
17	Wileca	0·6616
27	Jester	0·6584
3	Goodwin III	0·6159
24	Fione	0·6060

The handicap was based on the formula $\dfrac{\sqrt{L} + 2}{10}$ where L (the yacht's length) was as that used in the R.O.R.C. rating formula. The mathematics of the system seemed more difficult than in fact they were, and the result was the factors on the facing page.

To take *Voortekker* as an example: her elapsed time was twenty-six days, thirteen hours and forty-two minutes, her handicap factor was 0·8544, and therefore her handicap or corrected time was twenty-two days, sixteen hours and fifty-one minutes, which was the best time as it turned out, and Dalling won the Handicap Prize.

5
The start

The week before the start of the race was full of event, incident and extraordinary public interest. If there was a certain amount of ill-feeling about penalties and unfair advantages, it was more than compensated for by the round of social gatherings and the obvious enjoyment the competitors derived from them. They provided the opportunity for yachtsmen from nine different countries to meet and generate an unmistakable air of bonhomie and mutual respect.

The Royal Western Y.C. held a dinner for the competitors, their wives and a number of dignitaries on the Thursday. On the Friday evening, Plymouth Corporation gave a party for the competitors at the Council House. Four hundred guests were present, the Lord Mayor made a speech to which Follett replied, and the atmosphere was vaguely reminiscent of the Stock Exchange with last minute deals being made and information being extracted from competitors, sponsors and the Press. Non-yachtsmen huddled together in self-protective cocktail party groups.

The C. in C. Plymouth's sherry party on Friday had almost an Alice in Wonderland quality of unreality about it. An elegant house with a sort of 'Empire' atmosphere, immaculately dressed Naval officers being charmingly hospitable, and a collection of competitors who had been dragged off their yachts with grimy hands, grubby clothes and rather preoccupied expressions. Tabarly, most tired of all, strong and remote, was impressive because of his very stillness and his obvious concern about his boat.

Pen Duick IV was the centre of interest, especially on the

Robert Wingate, at 23 one of the youngest competitors in the Race, sailed Zeevalk. Voortrekker *and* Spirit of Cutty Sark *are 1968 successors of this 1949-built veteran, all designed by van de Stadt.*

The revolutionary proa Cheers *being put through her paces by Tom Follett, ex–World War II fighter pilot.*

Saturday morning as the yachts were towed out to the start. The brilliant sunshine of Friday had given way to a dismal, persistent fine drizzle which by mid-morning had turned to heavy, driving rain. There was hardly a breath of wind.

The Royal Navy's organisation of the start was stunningly efficient. In 1964, with only fourteen boats at the start, there had been a number of tangles and Howells' *Akka* was rammed by a spectator's motor-boat. Now, with thirty-four starters (Carozzo was still working on *San Giorgio* at 11 a.m.) there were no accidents and the competitors got away on the stroke of the hour.

That there was no wind to blow the yachts across one another's bows may have seemed a blessing. But without a breeze, the danger was helpless drifting. *Ocean Highlander* floated abeam right on to *Voortrekker*, with Munro unable to control or arrest the drift: Dalling just managed to hold his boat back. *Spirit of Cutty Sark* was carried right under H.M.S. *Clarbeston*, but Williams avoided a collision and got his boat clear. The Navy kept the assembly area and the race area (Fig. II) clear of unwanted and potentially dangerous spectator craft, and the public responded well to the directives and the printed instructions that had been published by the Race Committee the previous week.

The dismal weather kept the public away, and on the Hoe and up at Ramscliff Point there were fewer spectators than had been expected. The French Press arrived in force nevertheless and were mostly accommodated on a French Naval frigate, F.S. *Malgache*, which was sent, I was told, on the instructions of President de Gaulle.

Once clear of the harbour, the competitors encountered a very light north-westerly wind, and the early leaders were Tabarly, de Kat, Howell, Wingate, Gliksman, Follett and Bevan.

By 2.15 p.m., there was a flat calm and the sun came out. Most of the competitors were becalmed off the Cornish

Fig. III The Six Types of Rig used in the Race

The Chinese lug, seen in the 1964 race on Illala and Jester, was only used on Jester in the 1968 race. It is easy to reef, but a large spread of canvas cannot readily be set. The cutter has two foresails, and the photograph of Myth of Malham shows this rig to advantage. The sloop has one giant foresail which is harder to manage on its own. Silvia II and Spirit of Cutty Sark's pictures show sloop rigs set in action. The schooner has two masts with the front one smaller than the aft. Cheers and White Ghost were both schooners. Yawls and ketches also have two masts, and in both cases the aft (mizzen) mast is shorter than the front (main). The exact difference between a ketch and a yawl is obscure. If the rudder post, tiller and helmsman are all between the two masts, the yacht is a yawl. If they are all behind both masts, then she is a ketch. If the rudder post is aft of the masts, but the yacht is steered between the masts, then it is a matter of opinion what the boat is.

66

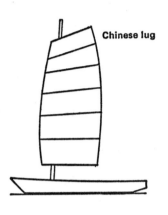

Chinese lug

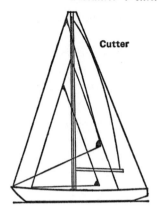

Cutter

Sloop

Schooner

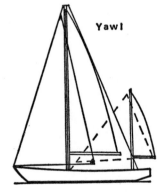

Yawl

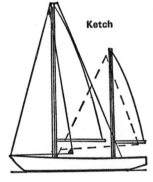

Ketch

coast – they wrote in their logs bitterly of the glassy seas and had time to look around and see where their rivals were.

Later in the evening, with the sky a brilliant blue and a heat haze over the sea, the fleet was some twenty-five miles out of Plymouth, between the Eddystone Lighthouse and The Lizard. Tabarly was still in the lead, coaxing and trimming *Pen Duick* to keep some way on her in the flat calm. Only a few hundred yards away the bright yellow of *Cheers'* paint glowed on the water. Slightly astern, *San Giorgio* had appeared from nowhere – having worked through the night and the morning Carozzo started two hours after the rest, ran aground in Plymouth Sound, and yet in just eight hours had caught up with the other competitors. Mysteriously, even after the Italian's appearance on the sea, there were still only thirty-four yachts making for America.

Edith Baumann was back in Plymouth, tearful and downcast. In conditions that were difficult for sailors who had much more experience, she had almost run aground on the Breakwater and with no wind to assist her she turned back towards The Hoe. In the evening, over a hearty meal of steak and salad with her parents and some friends, she brightened up, realised that all was not lost, and resolved to set out in pursuit of the others on Sunday.

In London, a vituperative telephone call to the offices of *The Observer*, gave some indication of the high moral tone that was going to be adopted by those who thought that the race was little more than a publicity stunt. The Editor of *Yachting World* had foreseen the hazards and answered many of the critics in that week's edition of his journal.

'A huge gust of fresh air has blown across the yachting scene. Rating rules, conventions – practically everything – has been thrown away in the interests of lightness, simplicity and speed. Among the entries are fifty-foot boats, weighing what some twenty-five foot boats do today: multihulls that, only a few years ago, would have been dismissed as

68

something from outer space. Where is all this going to lead?

'A certain element of eccentricity is bound to be present in an event of this sort, but the thing must be, and appears to be being, kept within bounds. By insisting on a five hundred mile qualifying cruise, the organisers automatically eliminate the majority of cranks who might otherwise be attracted. On the other hand, we must face the fact that, with growth as fast as we are seeing, trouble is inevitable.

'Whether it comes this time or next is unimportant; someone is going to be drowned. But to say something that can be said without offence before it happens, more easily than after, a death does not necessarily matter. It would matter a great deal if the organisation was at fault, or if the aims of the race were stupid, but this is not so.

'The challenge is exciting, and the race is giving yacht designers a whole new chapter to write. The Singlehanded Transatlantic is a "special case" and so long as the organisation remains in good hands, a death when it comes, and the resultant hullabaloo, must not be allowed to distort the overall picture.'

The experts knew the dangers and the competitors were aware of the risks. Gales, storms, flat calms, icebergs and fog are all hazards of the North Atlantic, and the yachtsmen no doubt bore them in mind when deciding which route to follow (Fig. IV).

The solo sailor faces a problem many months before ever he embarks on the race. The choice of route is of primary importance, as Tabarly showed when he devoted so much attention and study to it in 1964. In theory, there are five possible routes.

The Northern Route (estimated to be 3,130 miles) was taken by Hasler in both the 1960 and 1964 races. This part of the Atlantic is full of depressions, moving across from Canada to Ireland, and the depressions are as continuous and

69

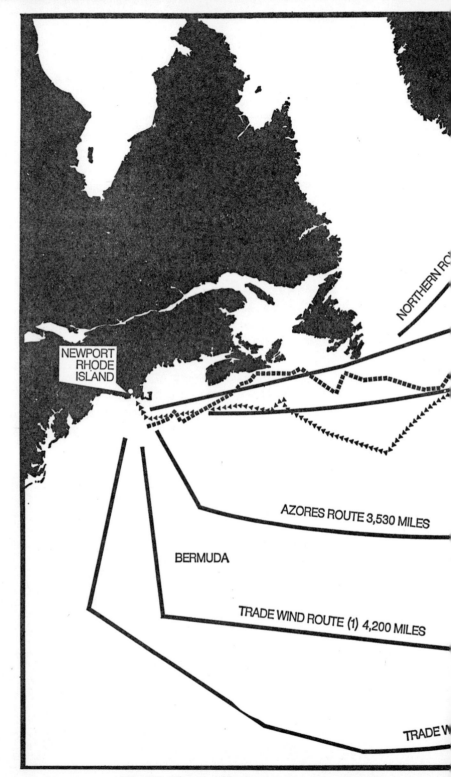

Fig. IV *The Possible Routes across the Atlantic*

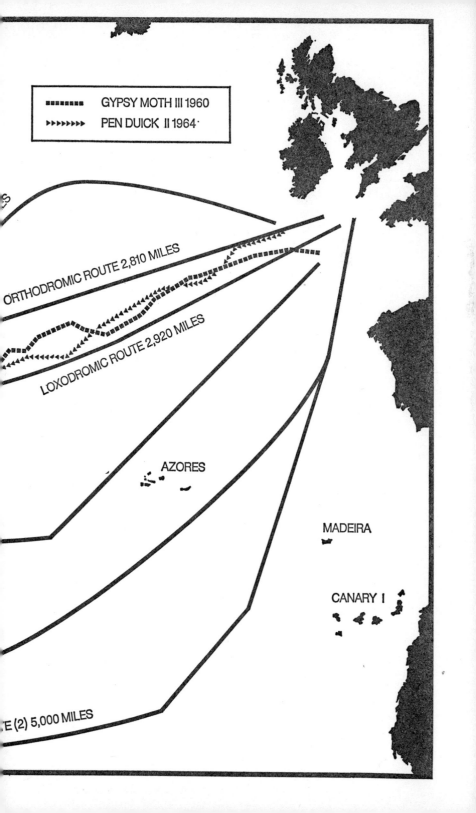

congested as traffic in Regent Street. Each depression is surrounded by a circular wind pattern, swirling in an anticlockwise direction. South of a depression, westerly winds are encountered (the yachtsman must sail into the wind). To the north of them, the yachtsman encounters favourable following easterly winds. Hasler sought these following breezes, but because he sailed so far north had to travel farther, and was also in danger of meeting the icebergs drifting down from the Arctic and the dreaded fogs off Newfoundland.

The Orthodromic or Great Circle Route (2,810 miles) is the shortest. Contrary to popular belief, this route is not a straight line from Plymouth to Newport (the straight line drawn on a globe passes across the terra firma of Newfoundland). A navigator taking this route has to draw two Orthodromic straight lines on his chart – from Plymouth to Cape Cod, and from Newfoundland to the Nantucket Light (some hundred miles from the finish). The sailor who elects to follow this route can be sure of encountering a continuous succession of head winds all the way across the Atlantic (though this was not wholly so during the 1964 race, when the winds were at times quite favourable) and also the west to east flowing Atlantic Current, both of which factors make the course an uphill sail. In spite of the disadvantages of the route, both Chichester in 1960 and Tabarly in 1964 (see Fig. IV) took the Orthodromic Route, though Tabarly was driven rather far south.

The Loxodromic or Rhumb Line Route (2,920 miles) is something of a red herring, because it appears shorter than the Orthodromic, on Mercator projection, the ones most commonly in use; where the Loxodromic line would seem to be the straight one between Plymouth and Nantucket, while the Orthodromic curves to the north (hence its alternative name of Great Circle). In the maps and charts reproduced here, the Gnomonic projection has been used. With this projection, any line drawn between two points is the shortest distance between those points, and it will be seen that in fact the Orthodromic Route is shorter than the Loxodromic. The Loxodromic's claim to attention is that it is the lazy man's dream – he can set his automatic pilot (a fine piece of equipment on a motor cruiser, and the use of which is not allowed in this race) to steer 260° True from Plymouth, and in no time, with a minimum of bother, he finds himself in Newport having sailed the Loxodromic Route marked on the map.

The Azores Route (3,530 miles) passes south of the depressions and the headwinds, and navigators taking the route can expect to enjoy a fair proportion of following winds, or at least winds on the beam – conditions in which multihulls might be expected to excel. Unfortunately, calms are a regular frustration, while sometimes the seas on the route are whipped up by the tail ends of hurricanes.

The path of the Trade Wind Route (anything from 4,200 to 5,000 miles) is one of controversy. The route marked 'Trade Wind Route I' on the chart is the one designated 'Trade-winds Route' by Tabarly in his book, and subsequently offered as the most southerly possible course. Yet the line of this route passes to the north of most of the trade winds. The winds will only be favourable to a yacht which sails far south, and 'Trade Wind Route II' – which Hasler suggested to me, and which he calls the 'proper trade wind route' – should find strong winds and few calms. It is a distance, though, of at least five thousand miles. In the two previous races, no one attempted to sail either route, or even came anywhere near so far south.

The other considerations in choosing a course for the race are wind conditions, fog, temperature and ocean currents at the time of sailing. There are charts showing these variable factors for each month of the year, and nearly all the competitors would have studied them. They provide an interesting pattern for the month of June (Fig. V).

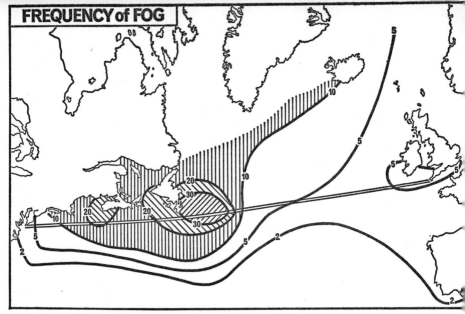

FREQUENCY of FOG

Fig. V

These maps show the weather conditions that can be expected for the month of June in the North Atlantic. Fog clings round Newfoundland, and the figures show the percentage of time when visibility is reduced to less than half a mile. Ocean currents are mostly half a knot, though they treble near the American coast. Solid arrows indicate a current persisting in that direction for 25 per cent of the time, or more. Dashed arrows indicate a persistence of under 25 per cent of the time. The sea temperature and iceberg limit chart explains itself. In June the break-up of the large bergs is well under way, and they drift ominously south into the path of the singlehanded yachtsman. The wind chart is the least reliable, for the wind has a law

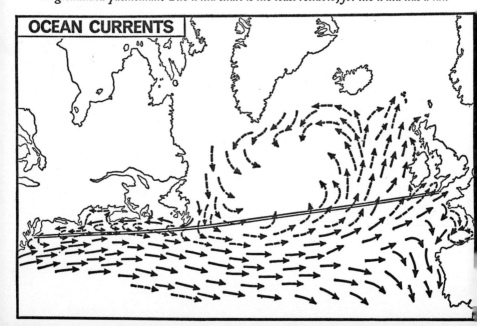

OCEAN CURRENTS

SEA TEMPERATURES

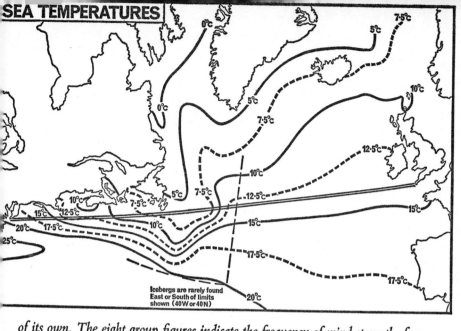

of its own. *The eight group figures indicate the frequency of wind strengths from force 2 to 9. Thus a group of 9, 19, 27, 15, 5, 2 + shows that force 2 winds persist for 9 per cent, force 4 for 27 per cent. The + indicates that force 9 lasts for less than ½ per cent of the time. Many 1968 solo-sailors wish this had been so. 'Dash' arrows indicate a 41–60 per cent constancy of the predominant wind. A 'dotted' arrow shows a 21–40 per cent constant direction. 'Crossed' arrows show two quadrants with similar wind frequency, while small arrows show that a wind frequently blows from another direction. The multiplication mark indicates no quadrant with as much as 21 per cent of force 3 winds or more.*

WINDS

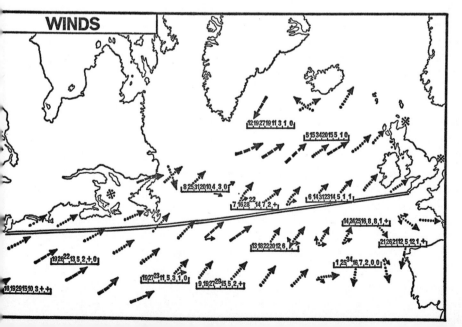

Most of the yachtsmen were cagey before the start about saying which route they would follow. An exception was Geoffrey Williams, who said that he would sail *Lipton* on the course recommended by his computer. When the news broke that he was to be assisted in his course plotting by an English Electric computer, the established yachting world virtually exploded. Red and furious faces appeared like a rash in the bars of yacht clubs, and many pink gins were no doubt indignantly spilt. The general feeling seemed to be that a computer was the last straw, and that if the cad refused to 'sail' his boat then the best thing that could happen was for *Lipton* to be banned from the race.

Much of the wrath was simply outraged conservatism, though there was genuine anger, and a number of letters were written to newspapers about the proposed violation of the ocean. Many of the arguments were confused, with fact doing little to dispel fiction.

During *Lipton*'s sea trials, Williams observed the yacht's performance and gave details of her behaviour to the Bureau Division of English Electric Computers. These performance codes were stored in the memory bank of a KDF 9 computer. Once he was in the Atlantic, Williams was to report his daily position to the Division, and this together with the weather information for the area where he was sailing, were to be fed into the computer.

English Electric claimed that they would then be able to simulate the progress of the yacht, and that the computer would advise on the best three of a possible five hundred routes. Williams, of course, being the only man on the spot, would still have to choose which one to follow. He did not, contrary to at least one public belief, have a computer on board. Though he did have Loran (a highly sophisticated and expensive form of navigational equipment by which a yachtsman can beam onto radio beacons on shore – in this case the coast of America – and by twiddling two or three knobs can pinpoint his position to within a few hundred yards). If anything gave Williams an advantage, his Loran set gave him a much greater one than did the computer. Loran is banned in all recognised offshore yacht races throughout the world, and a competitor was disqualified from the Bermuda Race recently for using such equipment.

Edith Baumann, still in Plymouth on the morning of Sunday 2 June, had no elaborate equipment aboard *Koala III*.

All she had were some pages from a nautical almanac, a few – very few – formulae, and a sextant with which she had been practising assiduously taking sun-sights – just the basic navigation she would need to take her to Newport. At lunchtime on Sunday she set out again, and with her sails nicely set, *Koala III* disappeared over the horizon some twenty-five hours behind the others.

The real drama of Sunday had been enacted in the early hours of the morning about fifty miles off the Isles of Scilly. *Pen Duick IV* had been racing along at a fair speed, narrowly missing a collision with a cargo boat in the middle of the night, when at 3 a.m. Tabarly decided to go below and make some coffee. He checked to ensure that there were no ships in the vicinity before descending to his galley. He had hardly been away from the deck a quarter of an hour when the whole yacht shook with the impact of a collision.

'There was a tremendous crash . . . it was a small freighter and it was anchored when I ran into it at about fifteen knots. I do not even know what nationality it was.'

A four foot gash was ripped in *Pen Duick*'s starboard float, the rigging on the mizzen mast was damaged and the mizzen crosstrees were bent. Tabarly brought the boat round and headed back for Plymouth. She limped into the Sound early on Monday morning, a disconsolate skipper at her helm.

Volunteers from the Royal Naval Engineering College, Manadon, and workers from Mashford's, spent all Monday repairing the float and straightening the crosstrees, and

Tabarly was almost ready to sail in the evening when it was noticed that the rigging on the main mast was also damaged.

'I will sail tomorrow, but I don't know whether I can make up three days. There are some very good boats in the race.' Tabarly was obviously disappointed. He had really thought that he could catch the others up after the initial setback, but the thought of unstepping the mainmast had clearly dashed his hopes.

The volunteers worked all night and through Tuesday to get the French boat back in the race, and Tabarly eventually set sail again on Tuesday evening. The trimaran disappeared out of sight at a cracking pace, though everyone in Plymouth knew that she had little or no chance of winning now.

Tabarly was hardly in the Channel when the rudder loosened and almost came adrift. He put into Newlyn overnight for repairs, but finally rejoined the race on Wednesday morning. The final blow came forty miles west of the Scillies when he found that his self-steering gear vibrated so badly when *Pen Duick* was doing more than nine knots that it shook itself to pieces. Back in Newlyn he said 'It is the same gear as I had on *Pen Duick II* [a Gianoli] but then I didn't go so fast.'

After leaving Land's End, he had reached a speed of thirteen knots in a force five breeze, and had been keeping pace with a distant motorboat for some hours, when the gear finally ceased to function.

'If only I had had a month's trial with the boat, it would have been a different story. The hull of *Pen Duick IV* is fast and solid. If I had had that collision in a wooden boat, I would never have made it back to Plymouth alone. Perhaps I will try again in four years' time – maybe in this boat. She could have won.'

He was disheartened and also heavily in debt. His immediate plans were vague, and he waited patiently in the dull Cornish drizzle for a friend to arrive from France and help him sail back to Lorient.

'Some people in France want me to do the crossing just the same. For me it would not have the interest of the race. I could try to beat the time of the eventual winner, but since we would not be meeting the same conditions it would be meaningless.

'On the other hand, it is essential that I sell the boat well and there is more interest in trimarans in the States. We will have to wait and see.' Tabarly was as philosophical as a man could be after so many frustrating setbacks.

Disaster also struck Hienemann on the Sunday when the self-steering gear on *Aye-Aye* broke. He sailed into Falmouth on Monday and the following day announced his retirement from the race. Four other yachts were forced to turn back and Dalling radioed that he was in some difficulty.

Lulled by the gentle zephyrs which hardly puffed their boats out of Plymouth Harbour, the lone yachtsmen may well have been caught unawares by the heavy seas and near gale-force winds they encountered on the Monday. On Tuesday, Dalling called me by radio telephone to tell me that he had lost his mainsail boom overboard. The unmistakable South African voice crackled across three hundred miles of Atlantic – 'This is *Voortrekker* calling.

'I have lost about six or seven hours' sailing in trying to recover my mainsail and boom . . . the pin in the gooseneck snapped . . . but I now have it jury-rigged and am going like the clappers again.'

Before this trouble, it had been thought that Dalling was well in the lead and he was obviously worried about the loss of time while he hauled his sail inboard again and effected the necessary repairs. The gooseneck is the pin and socket joint by which the boom is attached to the mast, and when the pin snapped, the boom went adrift, dragging the sail with it. Dalling must have had a job retrieving them, *Voortrekker* wallowing helplessly in the heavy swell while he worked.

No sooner had he fixed the joint with a makeshift pin than three of his lightweight winches broke. When I asked him how he felt about that, Dalling replied, 'It only hurts when I laugh, John, only when I laugh.'

Both *Ocean Highlander* and *Silvia II* were dismasted in the rough conditions, and Munro and Foezon headed back for Plymouth under jury rigs. *Zeevalk* succumbed to self-steering fatigue, and Wingate feared that an additional problem was her leaky hull.

'All my sugar and rice was swamped in gallons of water constantly sloshing about. But I have to go on.' Wingate's mother and sister were waiting to welcome him in America, having flown over (slightly prematurely) for the finish.

A still buoyant de Kat had put into Alderney to repair *Yaksha*, whose mast collar had been damaged and come loose. He intended to go on although the rigging was so loose that the stays resembled clothes lines flapping in the breeze. Tabarly had said in Newlyn that he was extremely concerned about de Kat, 'His boat isn't up to it and he has no idea what he faces out there.'

Meanwhile, the leaders were fanned out across the North Atlantic, several of them reporting their positions, though never at even approximately the same time – so comparisons were inaccurate and generally misleading. *Lipton*'s sponsors, who heard from Williams every day and twice a day, insisted that their man was out in front. I was not so sure.

Dalling called me again on Friday 7 June, to give me his position. After his setback, he had caught up with the leaders again. I gave him the positions of some of the other boats.

'If these positions are correct, it looks as though we are doing quite well . . . the wind is gusting about force six or seven . . . *Voortrekker* is making good speed, though I wish we could go faster . . . I am spending far too much time mending things and too little sailing.'

Blondie Hasler's creation, now owned by ace-navigator Michael Richie, Jester *sports a Chinese lug rig. She is the only yacht to have sailed in all three singlehanded Races.*

Designer Derek Kelsall and Martin Minter-Kemp won the 1966 Round Britain Race in Toria, so establishing multihulls in offshore racing. Under her new name Gancia Girl she led the trimarans into Newport.

Spirit of Cutty Sark was also up amongst the first few, though Williams was still having trouble with his right elbow.

'It is very awkward, John, and I can only work in short bursts . . . I've had to change my sail pattern . . . I now set a fairly large headsail which I leave up except in really bad stuff . . . I do all my reefing by rolling the mainsail with my left hand just steadying myself with my right.'

Another caller on Friday evening was Minter-Kemp. He had been easing *Gancia Girl* along in variable winds, and complained that breezes of constantly changing strengths were gusting from all points.

'I have not seen another boat since I left Land's End . . . just seagulls . . . for the first two or three days I was really frightened . . . I was seasick and could not eat or sleep. But I am settling down now and I drink a bottle of champagne for every ten degrees I sail west.'

He asked about the other boats, and from the position he gave me seemed to be lying fifth or sixth of those whose positions I knew. The news cheered him up.

'You know, when I look at my charts and see how far I have to go, I am absolutely terrified . . . but then I look again and see how far I have come, and I feel elated . . . this is tremendous sailing out here and a lot different from being in the Channel.'

During the week I had also expected to hear from *Raph*, but Gliksman had not even been in touch with his own magazine. However, the French boat was sighted by a long range reconnaissance plane of the French Air Force, which reported *Raph* to be neck and neck with *Lipton*, though slightly farther south. *Raph* was not sighted again, nor was Gliksman to report his position, until he reached St John's, Newfoundland.

Golden Cockerel was making good progress and Howell

seemed to be maintaining a champion rate of beer consumption. He was well up, though a good deal behind *Voortrekker* on much the same course.

By the end of the first week, *Pen Duick IV*, *Ocean Highlander*, *Aye-Aye* and *Tamoure* had retired. Waquet had returned to France with his strange little trimaran because of the strike by Air France pilots – which was at least an unique reason. The man who before the race had said that there was nothing to navigating the Atlantic then revealed that he had intended to check his position from the flights of Air France planes to and fro across the ocean. The strike wrecked all his plans for a fast crossing in *Tamoure*.

Zeevalk, *Silvia II* and *White Ghost* (Pulsford was having trouble with one of his floats) had returned to port, and eleven yachts – including *Cheers* and *Jester* – had not been sighted or reported their positions.

Of the remainder, five boats were contesting the lead. My entirely unofficial calculations, based on the latest known positions, placed the yachts in the following probable order, though they were all very close – *Voortrekker*, *Raph*, *Sir Thomas Lipton*, *Spirit of Cutty Sark* and *Golden Cockerel*, with *Gancia Girl* and *Myth of Malham* in sixth and seventh places respectively.

6

And then there were few

The pattern of the race may have emerged, but the excitement and drama had hardly begun. Gales and Mayday calls were now to play their part, together with a disturbing radio silence from competitors who were expected to call, and a frustrating dearth of news about some of the boats.

'Typical Atlantic weather,' was how Williams described it during one of his messages from *Lipton*. But he was lucky, and was to be routed, by the computer, north of the worst of the great storm.

The leading yachts sailed right into the depression on 11 June. The weather forecast that morning had warned of severe force nine winds becoming south-west force eight later. In fact the storm increased to force ten and at its height was probably force eleven. The Almanac has the following entry under 'Beaufort Wind Scale. Force 10. Limits of wind speed in knots – 48–55. Descriptive term – storm. Probable height of waves in feet – 29. Sea criterion – Very high waves with long overhanging crests. The resulting foam in great patches is blown in dense white streaks along the direction of the wind. On the whole the surface of the sea takes a white appearance. The tumbling of the sea becomes heavy and shock-like. Visibility affected.'

Such is the discreetly understated criterion. Dalling's log and the testimony of other competitors were altogether more emotive in their descriptions of a force ten storm.

The tired, almost despairing voice of Leslie Williams mumbled into my telephone receiver 'The storm was terrifying.' Williams had been having trouble with his radio, and the reception was very bad when he called from *Cutty Sark* some two thousand miles out at sea.

'That storm has cost me the race. I tried to get north of the centre to find the easterly winds, but I was caught in the eye of the storm. The indications are . . .' and his voice faded completely. At Baldock receiving station they fought to restore contact, and about a minute later he was on the air again.

'. . . vane was . . . my wind vane was blown right away by the high winds . . . just blown overboard. But I managed to rig a new vane from bits of plywood and attach it to the self-steering gear.'

Opus, Myth of Malham, Gancia Girl, Raph and *Golden Cockerel* were also caught in the storm. Howell said that the chaos in the catamaran was unbelievable, and that during the storm he had forgotten all about the race.

'I just concentrated on survival and drank beer to steady my nerves.'

Cooke radioed that after the storm he was feeling very cold, 'though the weather is now rather gentlemanly'. He also was caught by the depression and his barometer rose twenty-seven millibars in twelve hours.

'*Opus* suffered a knockdown when we fell sideways off a monster wave crest. Everything was a horrible mess down below, but on deck the mast, sails and steering gear survived.'

'Sailing through that was a nightmare,' groaned Minter-Kemp. 'That storm was terrible. I've never seen such seas. The waves were so high that it was like sailing the boat over the South Downs.'

Those yachts which were not in the leading group were more fortunate and missed the worst of the storm – although the depression continued to move eastward through the fleet before it spent itself near the coast of Ireland.

Gunthur III had missed it altogether, having returned to

Plymouth on 13 June with a broken mast step fitting. Piazzini explained the mystery of Gliksman's radio silence.

Gliksman had two radio transmitters on board *Raph* – a high-powered one, and a smaller set. Apparently sea water had penetrated the larger radio early in the race, and Gliksman had tried to use the smaller to make contact. But the signal was too weak, and the only person who picked up the message was Piazzini.

On 12 June *Ambrima* had been dismasted, and was taken in tow by a Spanish ship. The news did not reach the race office until much later in the month that although Cuiklinski was saved, the yacht had sunk, still under tow, off the Spanish coast.

Zeevalk left Plymouth again to try and rejoin the race, but the leak appeared to be getting worse and Wingate finally retired, putting into Penzance. *White Ghost* also made a second start, then a third and a fourth before Pulsford was forced to retire when the trimaran developed rudder trouble.

It might have been thought that a Swede would have been accustomed to the cold, but not Wallin. After sailing a northerly route for some days, he decided that the weather really was too inclement, and headed *Wileca* due south to the Azores in search of warmer climes. He said that he would sail the sloop across to the Caribbean where he could both charter the boat and be sure of some sunshine.

A troublesome cavity also played a part in the drama of the Atlantic, and Pyle gave 'crew toothache' as one of the official reasons for his retirement. He sailed *Atlantis III* into San Miguel with more than just pain – the self-steering gear had broken and some of the yacht's sails had been torn to shreds in the high winds.

Yaksha had set out again from the Channel Isles on 9 June and was reported to be taking the extreme Northern Route. Then on Tuesday 18 June at 7.30 a.m., Air Traffic Control at Prestwick reported 'a civilian aircraft has heard a distress frequency from trimaran number 43. Position 54° north, 30° west. Lost his rudder, lost his mast and lost one float. Unable to stay afloat. Abandoning ship. His last message.'

There immediately started the biggest rescue operation ever undertaken for a lone civilian sailor at sea. Shackletons at Ballykelly in Northern Ireland were brought to immediate readiness, and at 8.04 a.m. the first plane took off. From then on the rescue log reads like an extract from a wartime operation:

08.30: From Air Traffic Control Prestwick. Nearest ship is *Irish Rowan* in position 55° 54' north, 26° 6' west. Request H.M. Coastguard Bangor to ask *Irish Rowan* to assist. All details passed to coastguard.

08.45: From A.T.C. Prestwick. Atlantic rescue has ordered Ocean Weather Ship Charlie to proceed to north-east limit of its area and await instructions.

09.55: Request is made for Americans to supply an aircraft from their search and rescue base at Keflavik, Iceland.

10.00: S.S. *Alice Bowater* says she will be arriving in search area at 08.00 next morning.

10.35: Phoned *The Observer* in London for description of *Yaksha*.

10.43: Americans at Keflavik say a C97 Stratocruiser will be ready to take off at 12.30.

11.42: *Irish Rowan* asked to continue search. Continuous air cover is maintained over area from 13.00 hours.

And so on throughout the day. Two Shackletons and two American planes were airborne, combing the area. At least four merchant ships were headed there. The wind at sea level blew up to an awkward twenty knots and the cloud base descended to eight hundred feet.

It is worth remembering that the crews of Shackletons do not wear parachutes – they fly too low to use them. In the nine months before the de Kat rescue two Shackletons from Kinloss had been lost.

In the early hours of Wednesday morning a search plane reported that it was in trouble. Playmate 16 had been in the air for eight hours when her captain radioed that there was an engine failure, and that he was returning to the nearest base, Shannon.

Wednesday, 11.15: Playmate 16 lands safely at Shannon after the crew had jettisoned smoke flares and signalling equipment.

In the Kinloss crew room a pilot said 'If she had gone down, that would have been ten men lost for the sake of one.' The search continued throughout the day.

On Thursday morning the search area was moved to the south and east to allow for the effects of wind, current and drift. Several red plastic buoys were sighted in the area, but these did not come from *Yaksha*.

Thursday, 18.40: From Playmate 18, 'Survivor sighted 53° 55' north, 23° 45' west. He is in the dinghy waving and firing red flares.' Shackleton circles while request is made to New York coastguard for a surface picture.

21.00: Shackleton drops a Lindholme rescue dinghy with two survival packs containing food, water and cigarettes to the Frenchman, who climbs into the dinghy and appears to be reasonably fit.

The Shackleton also dropped sea-markers and radio buoys and continued to circle overhead. Two ships, both within a radius of thirty miles, answered the call and proceeded to the area. They were *Dutlon*, an American Navy auxiliary vessel with a civilian crew and *Jagona*, a Norwegian bulk carrier on her way from Canada to Latvia. The Norwegian vessel arrived first on the scene.

At 11.45 p.m. the *Jagona* radioed 'Survivor picked up' and finally reported 'de Kat on board and in good condition'.

De Kat owed his life to what the R.A.F. termed 'the eyeball, type human – Mark I' and a highly professional and co-ordinated rescue operation involving ships and planes of six nations. Coastal Command's Shackleton, Playmate 18, was packed with the latest detection and surveillance equipment normally used by a secret development unit at Ballykelly. But it was a crew member, not radar, who picked out the orange dinghy as it was buffeted and tossed in the Atlantic swell. Playmate 18 was not in fact searching at the time, but was on its way to a new area a hundred and fifty miles away. As the crew said, de Kat 'was a damn lucky man to get away with it'.

The harbingers of disaster were jubilant. Their direst prognostications had come true. Both the Royal Western Yacht Club and *The Observer* were accused of gross irresponsibility, of being careless of human life, of being too liberal and of trying to grab cheap publicity at the risk of human lives. The de Kat rescue had featured prominently as front page news in most papers for three days, and a surprisingly high moral tone was adopted in some extremely unlikely quarters. A favourite carp was the cost of the rescue operation, estimated variously (and always, it was claimed, accurately) at anything from £15,000 to £70,000 – and, the stories read, the bill must be footed by the poor taxpayer.

By far the worst piece of journalism was the television interview when the unfortunate de Kat, who was clearly genuinely grateful to his rescuers, and who had been speechless with emotion when he had tried to thank the R.A.F. officers in person, was asked by an interviewer whether he thought it was worth £70,000 to have his life saved. The Frenchman looked down, and apparently later wept – the interviewer seemed satisfied with his outrageous cross-examination.

The R.A.F. were not party to the emotional witch-hunt,

and behaved with great dignity. A spokesman said that it was impossible to gauge the cost of the rescue operation because many of the planes would have been on other sorties anyway, and added that emergencies of this kind were valuable training exercises. Squadron Leader Bullock, the pilot of Playmate 18, dismissed the question of cost. 'How do you put a price on a human life?'

The race had moments of great drama, but also it presented scenes of light comedy. Minter-Kemp told the story.

'On a brilliant calm morning a most astonishing stern trawler thundered over the south-east horizon and went round me in ever decreasing circles while her Gilbert and Sullivan crew addressed me in what might have been Serbo-Croat. My reactions must have appeared friendly, as the next move was a large steel bolt flung at me with considerable force and trailing a line. I hauled this in and discovered on the end, inside an unspeakable sock, a bottle of dubious brandy. Here was a dilemma; acceptance was obviously mandatory and yet could disqualify me. An exchange seemed an honourable way out and I flung a very small bottle of champagne on to a pile of nets in the trawler's stern. This was highly approved and the trawler departed over the north-west horizon to the strains of the Volga Boatman? leaving me wreathed in an aroma of very stale fish.

'About this time, I came on deck one night to find a wall of lights about a cable away to starboard shining through the fog. For one terrible moment I thought it was Sable Island – my *bête noire*. Then I realised that a very large liner had stopped and was flashing me to see whether I was all right. I rushed for the Aldis, and the liner responded by illuminating her vast funnels, which I found immensely moving. She slipped away into the mist leaving me rather forlorn and very damp.'

Contact with Williams aboard *Spirit of Cutty Sark* became more and more difficult – 'my radio is still very dodgy'. On the last occasion when Baldock managed to raise him, I heard the low wailing of a foghorn amidst the crackling and atmospheric interference. Williams explained the noise.

'There's a little ship – a coaster – come to look at me now, and he's hooting wildly. I cannot see the name from in here, but I'll go up and have a look . . . It's the *P. M. Crosbie*. I don't know what nationality, but that's the name. They seem to be wondering what I'm doing out here – but I'm glad to see them.'

The foghorn moaned again, and I, sitting in my office in the City of London, found the noise deeply moving. Williams signed off with 'See you in Newport, in about two weeks.'

On 19 June I left for America, and was joined in Newport by Chris Smith, who was to take photographs of the finish for *The Observer*. The momentary hiatus in news of the race was filled by the frantic activity and events in Rhode Island where over a hundred boats were preparing for the Bermuda Race which was due to start on Saturday.

There were still no reports of the whereabouts of *Cheers*, *Voortrekker* and *Raph*. Then on 21 June a telephone call from Gliksman to me in Newport told of his ill-fortune. *Raph* had a damaged rudder, and the Frenchman had put into St John's, Newfoundland, for repairs. He was bitterly disappointed, and although he hoped to set out again after the week-end, he realised that his chances of winning the race had disappeared.

Earlier that week, there had been a report of a two-masted yacht off Sable Island. The name was not seen, and the sighting caused considerable puzzlement in the race office. It could not have been *Lipton*, whose position was known to be some hundred miles farther out in the Atlantic. I had assumed that the boat was *Voortrekker* – but after talking to Gliksman when he came to Newport later in the month, I could only conclude that the phantom yacht was *Raph*, and that he was in the lead before his rudder broke.

Raph set out again on 25 June, but 'the boat was not right' and Gliksman returned to San Pierre where he retired from the race.

'I sensed that my luck had run out,' he explained. 'My main fear was being dismasted in mid-ocean and being left without a radio. Quite often I was going very well to windward and then I would let in my sails for fear of being dismasted. The main trouble was the sails – they are too heavy and I could not have sustained a tacking duel. It took me a long time to tack.'

La Delirante was spotted returning to France with a broken mast on 23 June, and *San Giorgio* limped into Falmouth with a damaged rudder four days later. Carozzo was chagrined that a large piece of driftwood should have dashed his hopes of winning after all the trouble he had been to in getting his boat to Plymouth originally.

By the end of the third week, *Fione, Dogwatch, Jester* and *Cheers* still had not been sighted.

The radio station at Baldock continued to receive signals from *Voortrekker*, but they were so weak that it was impossible to make contact, and they got weaker, showing at least that Dalling was still sailing westwards.

With no information about *Voortrekker*'s position except the one Dalling had given during his last call on 7 June, it was generally assumed that *Sir Thomas Lipton* was in the lead, with *Myth of Malham* (Bevan must have been driving her like a speedboat), *Spirit of Cutty Sark, Golden Cockerel* and *Opus* all in contention for second place. *Maxine* was moving westward at a fair speed, and *Silvia II* (Foezon had set out from Plymouth with a new mast on 12 June) was screeching across the Atlantic.

Then, on Sunday 23 June, a plane of the Royal Canadian Air Force on a routine flight off Newfoundland reported seeing both *Sir Thomas Lipton* and *Voortrekker*. Williams was no more than sixty miles ahead of the South African. Newport held its breath as the race burst into life again, and the attention of the yachting world and of much of the general public was riveted on the five hundred mile last lap of the race.

But before any of the competitors was to reach the finish, a new drama unfolded itself in mid-Atlantic, three hundred miles north of the Azores.

Koala III had been making gentle if somewhat sedate progress towards the American coast for the first two weeks of the race. Edith Baumann said that everything had been going according to plan – 'I felt elated.' Then the wind began to freshen, and the seas increased – the tail-end of the storm that had battered the leaders now swirled about the trimaran. Without warning, one of the metal stays joining the port float to the main hull, snapped, and Edith Baumann realised that her boat was in trouble. She radioed Waquet to ask his advice, and he told her that the damage could be repaired and that she should alter course for the Azores. She headed for the safety of the islands, but the storm became worse. Huge waves smashed another float stay, and she sent out a routine call, not knowing who, if anyone, would hear her.

That was on the evening of 25 June. Later that night she saw a ship in the distance, and fired a red distress flare. She described what happened next.

'A very large freighter came to me at about two hundred yards, remained there one hour, left, came back when I fired another distress signal and then went away without assisting me. She was so close I could see the crew moving about on deck. Yet they left me all alone in the angry sea.'

With her port float practically severed and no response from her steering, she realised that *Koala III* was in considerable danger of breaking up, and Edith Baumann sent out a Mayday call on 26 June. Then she just waited and prayed.

The call was picked up, and a Hercules of the U.S. Air Force spotted *Koala* with her skipper and Schatz still on board – it had earlier been reported that she had abandoned ship. The Hercules continued to circle the yacht and three ships in the area steamed towards the rescue spot. The German freighter *Magdalene Vinnen* was first on the scene, but the sea was so rough that she could not get near the stricken yacht. Then the *Henri Poincaré* (a French research vessel) tried to take the yachtswoman off the trimaran, but the mountainous waves threw the yacht against the side of the ship, the mast came crashing down, and Edith Baumann prepared to jump.

The *Henri Poincaré* steamed away and hove to while her captain launched a dinghy. The small boat came alongside *Koala III* and the German secretary and her dog were taken off to the safety of the rescue ship.

It was not possible to pick up the trimaran and her skipper watched the little boat swept away and with it all her personal possessions – the party dress she had taken along to wear at the receptions in Newport and the log of her voyage.

Once again the witch-hunters came out in force, saying that a woman with so little experience should never have been allowed to enter the race. It was claimed that her trimaran was unseaworthy – yet when I had seen it in Millbay Dock it had seemed one of the strongest and best built multihulls in the race.

Fortunately, the sour taste left from reading the sensational and often ill-informed reports of the German girl's misadventure was removed by the dramatic and record-breaking arrival of the first yachts in Newport.

7
The finish

From Sunday 23 June onwards, the excitement, anticipation and speculation in Newport knew no bounds. Elderly American ladies kept asking me if it was true that *Lipton* was going to win, and when could they see the boat.

Some of them, I thought, must really have believed that the yacht was being sailed by Sir Thomas himself. They remembered him as 'a charmer – so very English, and such a gentleman', and they had seen him on the five occasions when he had brought his challengers across in all those vain efforts to wrest the America's Cup from the tight grasp of its holders. Now the old ladies did so want to see the fifty-seven foot ketch win.

Lipton's sponsors, and the representatives of the news-paper for which Williams had been writing, were confident of victory. I was not so sure. Brasher, in London, was passing on the latest reports of sightings to Hasler, who was plotting them and comparing the positions with Tabarly's in 1964. On Tuesday, I heard from Hasler that the Sunday positions of *Sir Thomas Lipton* and of *Voortrekker* made the English boat four hundred and eighty miles from Newport, and the South African five hundred and forty. Sixty miles between them and over four hundred to go. This, I felt certain, was going to be a race to the finish and if the boats were neck and neck, then I did not give much for Williams's chances. Dalling was much the more experienced sailor, I argued, and there were still considerable doubts about Williams's ability as a navigator. Navigation on the run-in could play a vital part; there are shoals and little islands dotted all about the coastal waters off New England.

Hasler's calculations showed *Lipton* to be still one hundred miles behind Tabarly's equivalent position in the previous race and it had taken the Frenchman five full days to sail in from that point. Nevertheless, Hasler calculated that Williams (or Dalling) could just conceivably arrive on the Wednesday night, though he said that he thought it unlikely that the record would be broken – unless *Cheers* slipped quietly in.

The *Cheers* 'project group' were beginning to look rather less confident. I think that Newick and Morris still believed that Follett would win, but they were clearly worried that he had not done so already. They were not in the least alarmed about the safety of the boat or her skipper. But they had set themselves a target in time for the crossing, and they had genuinely believed that Follett would be in on schedule over the weekend.

By Wednesday, the brilliant weather of the previous week had given way first to fog and then to rain. While Newick and Morris looked dejected and wet in their oilskins, the rest of us peered nervously through the mist of rain, wondering if the yellow hull of *Cheers* would appear round the headland like a bright light in the gathering gloom.

The tension was becoming unbearable, and not a few tempers were on edge. His sponsors had apparently lost contact with Williams, there had been no word of Dalling since the sighting on Sunday, and as *Cheers* did not have a radio and had not been sighted since she left Plymouth, no one knew where to look for her.

Smith and I tried to charter an aeroplane to search for the three yachts. But air safety regulations stipulated that we could not fly out of sight of land, and close inshore was not where I expected to find the boats. The weather conditions were so bad that any sort of extensive search became impossible. Then, on Wednesday afternoon, the high powered radio receiver in Dunning's office at the Port o'Call crackled into life.

It was Williams calling from *Lipton*, and he was being picked up by the Nantucket Light Vessel, and monitored through the Castle Hill Coastguard Station.

'I can now see the Buzzard's Light, I can see the Light, and that is the first light, buoy or land I have sighted for twenty-six days.' The emotion in his voice was instantly communicated to all of us who were listening, and it was unforgettable to hear a man talking so after he has sailed three thousand lonely miles and realises that he has less than a hundred to go.

Then he gave his position, and Dunning plotted it on the large scale map. I realised immediately that Williams was sailing to the north of the Nantucket Light, and explained to the others that he was meant to sail round the lightship – in other words to the south – to comply with the rules of the race. All hell broke loose.

I telephoned Odling-Smee, who was at the Thomas house, and gave him Williams's reported position. Odling-Smee and Thomas translated the figures to their map and both realised that *Lipton* was cutting the corner and going north of the Light. Odling-Smee thereupon called Muessel at the Coastguard Station and asked him to inform Williams that according to the sailing instructions, competitors were supposed to sail south of the Light Vessel.

Time crept by, with *Lipton* doing an estimated speed of seven to eight knots, and Muessel called back to say that Williams claimed that his sailing instructions stated he should go south of Nantucket, and that was all – there was no mention of the Light.

Again Odling-Smee said that Williams should check, as all competitors had been told in Plymouth to amend their instructions.

Muessel spoke with Williams, who insisted that his

Alan Gliksman's aluminium Raph, *the longest monohull in the Race, was probably leading the fleet when her rudder disintegrated off Newfoundland.*

*Bearded Joan de Kat surveys the shoreline from his
frail and doomed trimaran* Yaksha, *as she lifts
her weather hull gingerly out of the water.*

instructions said nothing about the Light, and he asked the Coast Guard Commander to ascertain from Odling-Smee whether or not he should go about and sail round the Nantucket Light.

By now *Lipton* was due north of the Light, and right in the middle of the treacherous Nantucket Shoals. Odling-Smee was in a quandary. Was he to ask the skipper of a large ketch, drawing nearly eight feet of water, to come about in such turbulent waters, when the boat was doing a fair speed, in order to sail back and around the controversial Light. The Chairman of the Race Committee took, as he later described it, a very deep breath, and told Williams to follow his sailing instructions as they were written.

Odling-Smee said: 'To have asked him to come about there might have put him in some trouble, and possible danger. I was influenced by the fact that he was in these Shoals, with the wind rather on his quarter. I had to think very quickly while the poor fellow was waiting on the end of a blower. I was not being judicial or anything, but to tell him to alter course and sail round might have been lethal.'

So Williams held his course, and sailed on in. He estimated that he would arrive early on Thursday morning, but during the evening the wind blew up a knot or two, and it became apparent that Lipton would near Brenton Reef Tower in the middle of the night.

At this time I was still convinced that Dalling was in the lead. It is true that there had not been any sightings of *Voortrekker*, but then visibility was so bad that she could easily have slipped undetected past the Light. All through the evening I held to my view, as Williams sailed closer, his voice getting louder and stronger and more excited.

By midnight, Smith and I could bear the suspense no longer, and we took out a fast boat to search for *Voortrekker*. Smith did not agree that the South African boat was in the

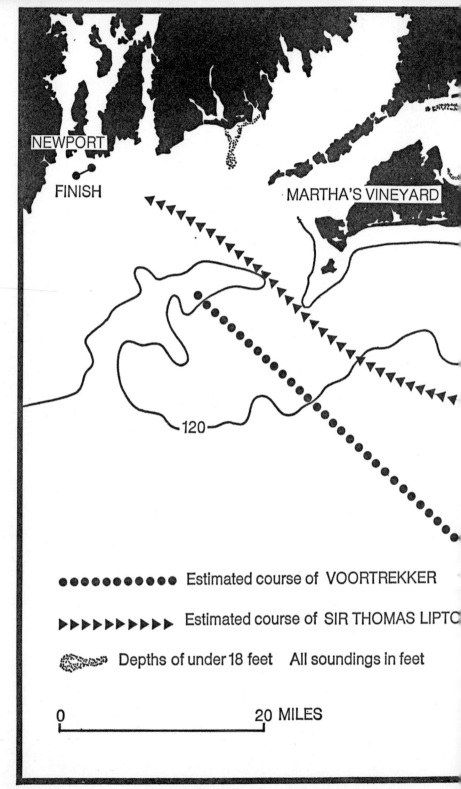

NEWPORT

FINISH

MARTHA'S VINEYARD

120

●●●●●●●●●● Estimated course of VOORTREKKER

▶▶▶▶▶▶▶▶▶▶ Estimated course of SIR THOMAS LIPTON

Depths of under 18 feet   All soundings in feet

0 _____ 20 MILES

Fig. VI The Approaches to Newport

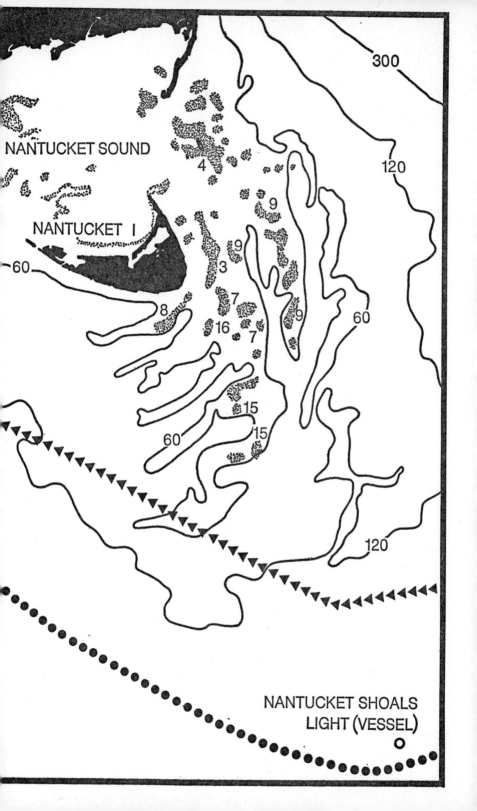

lead – he was only interested in photographing the winner. For two hours we searched the Atlantic beyond the finishing line and then we spotted a sailing boat on the horizon. She was a ketch, which did not help very much, as both *Lipton* and *Voortrekker* were rigged as ketches.

The sky was overcast, and there was a heavy sea running. There were no lights on the yacht (owing, I later learned, to wiring trouble) but the searchlights on the decks of the small flotilla of welcoming boats played on her sails and white hull. It was *Lipton*.

In the glare of the strong lights, I could see Williams standing in orange oilskins in the stern of his boat. I peered through my binoculars – he was crying.

Newport had planned a tremendous reception for the winner of the race – fireboats, hooters and throngs of people. But Williams made such good time from Nantucket that he arrived several hours earlier than he and we had estimated. And two-thirty in the morning is too early for fireboats. After we had found him about two miles from the finishing line, the ships sailed alongside, and had to put on a lot of power to maintain contact. *Lipton* was on a spanking reach, and looked quite magnificent as she slipped through the swell at an easy ten knots.

All the arguments and controversy of the past few hours were forgotten as we watched an extraordinary young man sail in at the end of a lonely voyage across the Atlantic. *Lipton* is an enormous boat for one man to handle, and Williams sailed her beautifully past Breton Reef Tower at 2.33 a.m. local time.

Williams hove to, and lowered his sails. The Coastguard boat went alongside, and made fast, and Odling-Smee and Thomas congratulated him on being first across the finishing line. Two boys who had been pupils of Williams's when he taught in New York jumped on board *Lipton*, and he was obviously delighted to see them.

He was towed into Newport to a cacophony of hooters and cheers from a quickly assembled and surprisingly large crowd. At the quayside he answered endless questions, posed for photographs and refused to catch up on his sleep. He worked on the boat, answered telephone calls and exhibited a boundless energy that astounded those of us who were wilting after our all-night vigil.

Later on Thursday morning, Williams showed me his sailing instructions, and there was the printing error with no amendment concerning Nantucket 'Light'. He admitted that he had been at the pre-race briefing, but could not recall being told to change his instructions. He said that he had been called away to answer a long-distance telephone call, and assumed that he must have missed the verbal alteration. Once again I thought what a pity it was that Williams had remained so aloof from his fellow competitors in Plymouth. Had he been part of the crowd he should no doubt have been told what he had missed at the briefing – and a good deal of embarrassment and unpleasantness would have been avoided.

Not until two-thirty in the afternoon could *Lipton* officially be proclaimed the winner, and waiting for the twelve hour penalty period to elapse was agonising – especially for Williams. Late in the morning, the Coastguard boat went out again to look for Dalling. Watchers on the Nantucket Light Vessel had reported seeing *Cheers* sail past the Light, and it seemed possible that the American boat would take second place. Dalling had to be out there, but without his radio to assist us, it was, in now very heavy seas, entirely a matter of chance whether or not we spotted him.

Aboard the Coastguard cutter were Muessel, who was in command of her, Odling-Smee, Thomas and myself. I had been taken along not because I was reporting the race for *The Observer*, but because I was also acting as the representative of the paper in Newport. Odling-Smee and Thomas discussed the problems of Williams's penalty and his failure to sail round the Light. They did not consult anybody, though they did ask me to listen to their deliberations, and to issue a statement on their behalf. Whatever decisions had to be made were Odling-Smee's responsibility, and this was clearly understood, though I did say that if there was any likelihood of disqualification, then the news should, on humanitarian grounds, be conveyed to Williams as soon as possible.

The Chairman of the Race Committee had already asked for, and received from Muessel, a sealed letter in which the Coast Guard Commander had written his estimate as to the time saved by Williams in sailing to the north of the Light; and Odling-Smee intended to use this as a guide in the event of a protest.

At three o'clock on Thursday afternoon, I issued the following statement on behalf of Odling-Smee. It was typed on paper headed '*The Observer* Press office', and said: 'The Race Committee has asked me to state that the winner of the 1968 *Observer* Singlehanded Transatlantic Race is Geoffrey Williams in *Sir Thomas Lipton*, subject to any possible protest.' Odling-Smee had made it clear that he would entertain a protest only from another competitor, within forty-eight hours of that competitor's arrival in Newport.

Williams at last retired to his hotel at about six in the evening, and missed the arrival, some seventeen hours after his own, of Dalling and *Voortrekker*.

Dalling crossed the finishing line at seven forty-two, local time, and was escorted in by the mayoral boat. He looked dejected and rather forlorn. As I stepped on board his yacht, Dalling looked sadly at me and said: 'Second? Second is no better than last.'

Fourteen and a half hours later Follett sailed *Cheers* past the Brenton Reef Light, and so won for himself the third prize, and also the prize for the first American in the race. *Cheers* had in fact reached the finish the night before, not very far behind *Voortrekker*, but as there were no boats about, and Follett was not familiar with the coastline, he hove to and rode out the night, finally sailing across at six thirteen on Friday morning.

Soon after breakfast, Dalling was down at the Port o'Call working on his boat, tidying, cleaning, stowing and above all drying things out. He seemed much calmer – understandably, after a sleep in a dry, firm bed – and was in greater control of his emotions than he had been the night before. He was bitterly disappointed at not winning – but there was no trace of self-pity. He simply felt that he had let down 'the people back home'. 'They just won't understand,' he kept repeating.

'The Race Committee asked me if I wanted to protest. It hadn't entered my head. If you win the race you are the winner, and that's it. You can't begrudge a man winning a race like that, it's simply impossible.'

And on the question of the computer he said: 'I don't think a computer made much difference. Geoff won because he has a better boat – *Lipton* must be faster as she is longer on the water line than mine. In that storm I read the weather signs correctly and tried to go north, but I worked right into the worst of it. A computer wouldn't have helped much then.'

That was not the opinion of some of Dalling's backers. The next forty-eight hours were unpleasant for us all – two days of rumour, counter rumour and vituperation. They were worst of all for Dalling, who throughout remained adamant that he would not protest.

'Under no circumstances will I consider a protest. It is up to me to protest, and if they [the sponsors] do, I will fight this thing all the way. They should get in touch with me, as I sailed this bloody boat and the decision is up to me.'

The storm in a mini-teacup blew up when a member of the Voortrekker Trust sent a cable to *The Observer* (and a copy of it to the Royal Western Y.C.) which read: 'I must protest against the result of the race, The *Thomas Lipton* having sailed against the rules. Have advised Dalling accordingly. Signed: Transfon, on behalf of the Voortrekker Trustees.'

Fortunately the chairman of the Trustees, Mr R. V. Norton, made it clear soon afterwards in South Africa that Mr Transfon was acting on his own initiative, much to the annoyance of his colleagues, and that the protest was strictly unofficial. Mr Norton said: 'We accept the decision announced by the Race Committee and we have sent a cable of congratulations to Geoffrey Williams.'

Dalling was obviously relieved to hear that. After twenty-six days of emotional and psychological hell in the middle of the Atlantic, he clearly did not want to have to play at international politics. And so Newport got down to the serious business of fêting the three yachtsmen, all of whom had broken Tabarly's 1964 record for the solo crossing from England to America.

Newport is very experienced in celebrating the triumphs of yachtsmen, and gave the three men little peace or time to relax.

They unwound, nevertheless, each in his own way. Williams wanted to talk, and did so at length when he was not being taken off to some official function. On the Friday morning, he was made an honorary citizen of Newport at an informal ceremony in the Mayor's office.

During the days immediately after his arrival, the young Cornishman seemed to need time neither for mental nor physical relaxation. Except when he was working on his boat, he was always immaculately dressed, an apparently smooth man who had just accomplished an extremely rugged feat. He quickly shaved off the rather straggly little beard he had grown during the voyage, and dressed in the best piece of blue suiting a well-known London boutique could put together for him, set out to charm Newport.

He wanted desperately to be liked. He realised what a bad press he had had before the race and how arrogant and unhelpful he had been accused of being. *The Observer* had not been impressed by his refusal to talk of himself or his boat, and I had said some fairly harsh things about him. Some of these I now had to retract, for Williams did not mean to be rude – that was obvious. He just seemed to have a very unfortunate manner, which, until you got to know him, set you rather on edge.

Williams is articulate and interesting to talk to – providing you know about boats, preferably his boat, and do not want to talk about much else. He is calm, with penetrating eyes and a huge handshake, and the occasional West Country burr comes through under his Oxford accent – generally when he is excited or tired.

In Newport he talked endlessly about his plans and ambitions, of his proposed sail training centre, of his intention to build twelve metres and win the America's Cup, of his decision to go into politics. All very cut and dried, and ambitious. But it did not sound especially arrogant the way he said it – more than anything he seemed slightly insecure. Or perhaps it is just that he is a loner.

He was not frightened by the Atlantic, 'the danger of the solo crossing is greatly overplayed'; nor was he ashamed of crying when he crossed the finishing line, 'it was a tremendous satisfaction'. When a local reporter asked him why he undertook the venture, he replied flippantly: 'My mother never gave me enough toys when I was young.'

Later, he admitted that the remark may be more profound than the Newport journalist realised, and he added: 'It was something to do with the fact that I never got a rugger Blue at Oxford. My father was very disappointed about that as well.' And yet I have been assured by members of Vincent's

Club, that Williams was not nearly good enough when he was at Teddy Hall to be awarded a Blue.

'I have always admired individuals who do things really well. That is why I wanted so much to win this race. I'd have swept the streets to make money for a boat to win the race in.' But he never had to – he worked as a master in a New York school where he tried to teach the boys to play cricket. One of his ex-pupils told me that he was not a very popular master, but that the boys who did like him, liked him a great deal.

There is this strange streak of T. E. Lawrence in Williams, and Williams admires Lawrence greatly. He read 'Seven Pillars of Wisdom' during the voyage, and clearly has the same capacity for commanding, and demanding, devotion.

But he was too unemotional for credibility about the dangers of his crossing. He expected me to believe that at no time was he scared, and that his only reaction after winning the race was that it was simply a means to an end.

'All this win means to me is that I'll be able to do the things I want that much more easily. It will help me with my sail training project.'

Follett, at fifty, is exactly twice Williams's age. He is bearded and weatherbeaten, but has an extremely kindly face, with a perpetual smile breaking from the corners of his mouth.

He arrived with much less hullabaloo than Williams or Dalling, and appeared more unemotional than either of them. But appearances in this case were, I am certain, deceptive. Follett is the complete professional – the very best sort of professional – a master of his craft, quiet and unassuming.

In Newport, Follett was even more charming and unassuming than he had been in Plymouth before the race. He was quite honest about coming third: 'I should liked to have won.

'But if I have any real disappointment, it is that I didn't make the voyage in the time I predicted – twenty-one days.'

He is the veteran of four singlehanded crossings of the Atlantic, and I asked him why he did it. 'It's just that I like travelling alone. You see, I'm a solitary. I am not really interested in racing, but much more in proving that *Cheers* is a fast boat.'

By taking a southern route, though not an extreme southerly Trade Wind route, he avoided the great storm that tossed some of the other competitors about. But even in the warm southerly climes, he encountered sudden squalls and one gale, when for seven hours he sailed under bare poles. His log entries were unemotional and seemed almost to play down his achievement in sailing the tiny little boat across. Nowhere in the log did he write at length of his bad luck, nor did he try to make what he recorded sound emotional. He described the gale as 'Very odd weather today.'

In a flat calm he displayed a wry humour, and the entry reads 'Flat calm again. Come on wind! You're costing me £500.'

Towards the end of his voyage he obviously began to get a trifle tetchy about the variable weather – where other boats can tack to meet changing winds, Follett had the business of bringing *Cheers* round, exhausting work even for such a fit and experienced man.

'Fluky wind. All sail off till it settles down. Looks like I'm trapped here 135 miles from the Nantucket Light.

'Thick fog. No wind. Oh joy! Wind shifting. When will it end? Close hauled on port tack. Here we go again! On starboard tack.

'Very lumpy sea tossing us about. Still, we make a little progress towards the Lightship. What dreary weather! Worse than England. Never have seen such fluky winds as we've been having today. Heavy swell rolling in from SW plus other swells from here and there. A very rough ride even though off the wind.'

The economy of the language is as crisp and workmanlike as the man himself. For the Folletts of the ocean, mental anguish is hard to put into words. It just grows, over the years, into the lines of the face and the toughness of the hands and the sensitive ability with which a boat is handled. Follett's final entry in the log was purely factual.

'11.20. Crossed finish line. Pilot boat came out. Coastguard came out with Dick Newick and Jim Morris and towed me to Port o'Call Marina. Came in third beaten by *Sir Thomas Lipton* and *Voortrekker*.'

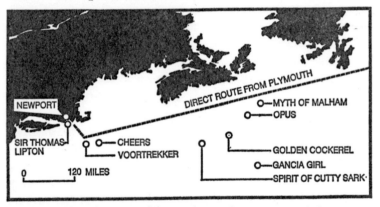

Fig. VII The Positions of the Leaders as the Winner Sailed in.

The horrendous implications of facing those three thousand miles of Atlantic, and the true measure of the men who did so, can be seen in the log that Dalling kept so meticulously throughout his voyage.

Dalling is thirty, with fair hair and a beard. His blue eyes smile and he is gentle. In Newport he made me deeply aware that he had just been through three and a bit weeks of psychological hell. His hands were calloused, his nails cruelly bitten. He gave the impression of being a perfectionist, a thinker and, transparently, a romantic. He is also one of the most articulate men I have ever met, which is why his log was so readable.

Page by page and day by day he set out his lonely fears and self-doubt. Later, when the reaction of having arrived safely had set in, he was a little ashamed of what he had written – he felt that no one reading the words on dry land could understand what it was all about. But perhaps understanding was easier than he thought.

On the first page of the log, before ever he had set out, Dalling wrote a quotation from Whitman – 'Do we think victory great? So it is, but now it seems to me that, when it cannot be helped, defeat is great, and death and dismay are great.' And then as he sailed down the English Channel, hardly a day out of Plymouth, he made this entry – 'I must be nuts' – and he meant it.

Because he writes so well and because he found time to think about his voyage as he sailed, I believe that Dalling's log best describes the emotional and psychological problems encountered by lone yachtsmen in mid-Atlantic. Leslie Williams kept a touching record of his trip in *Spirit of Cutty Sark*, and others wrote of their feelings in the storms and calms. Geoffrey Williams says that he did not write anything. But Dalling's prose is especially readable and moving.

During the gale so many of the competitors encountered half-way out, he and *Voortrekker* had their worst moments of the trip.

'She seemed quite comfortable under yankee and mizzen only, and made quite good speed. I marvel at how she holds together. I can never get used to bad weather – there is so little one can do after a while.

'I feel emotionally shattered by it all . . . at times like this I will be glad to get there in one piece.'

There was much worse to come. 'I write this on Wednesday 12th – I have had neither the time nor the energy to concentrate on writing. The gale moderated at about dusk, and left the boat soaking wet and in a shambles. The sun came out, and filled with that fantastic feeling of relief and reprieve that comes after a gale at sea, I cooked myself a meal of sausages, eggs and berries, and had a couple of whiskies. Absolutely out, I slumped into a sopping blanket and slept fitfully. The wind backed to south and then to east and then north-east, and I should have smelt a rat.

'But the thing I did notice and did not pay enough heed to was the drop in the barometer after getting to 29.78. I awoke at 0700 with the boat doing 9 knots and well heeled over . . . a glance at the barometer, and my heart nearly stopped . . . . the next twenty-four hours were the nightmare of a lifetime that will be the subject of my dreams and will no doubt crop up in my thoughts even while awake. By 1000 the seas were as big as the ones in the Mozambique Channel that did me in while sailing *Carina*.

'Now, I stood below steering with my legs and hanging on to the weather grip rail, peering at the seas through the doghouse window. I judged the position of the boat by watching the strakes of foam on the water. At its height, the wind speed was at least 50 knots. Three times, petrified, I had to leap out of an open hatch and lash things to the deck which had broken adrift. The height of the waves I conservatively estimated at 35 feet, but there were some exceptional ones that must have been all of fifty. I prayed very hard during this storm.

'We took some waves right over us – she would heel over to the horizontal and everything below would go a lovely deep blue colour. Then she would come up slowly, shaking the water from her. My mouth was permanently dry with fear and I nearly fainted a number of times from exhaustion. It started easing at about 2000, so I went on deck, tidied it all up and adjusted the self-steering. I think that was the most courageous thing I have ever done in my life. It might not seem like it later, but I know what it was like at the time.'

Dalling thought that his chances of winning had completely vanished when he wrote:

'There's no wind here, and I know that everyone else will have a good breeze closer inshore . . . becalmed! This is really chronic. Not a breath. Glassy sea. I am going through mental agony at present – flaying myself for this lack of progress. This business of being at the helm the whole time is a killer. I feel so despondent I can hardly write about it. Everything has gone wrong in the last week. To say that you have been becalmed for four days when you are within spitting distance of home sounds awfully pathetic.

'Have just heard that Joan de Kat has been sighted by rescue aircraft – three cheers.

'But it seems as if I'm going backwards. I've not even got any appetite any more. I wish I had brought the Book of Common Prayer – it has some beautiful things in it. I wish for an end to this ordeal now – I'm absolutely shattered in mind and spirit. Only thoughts of those at home are keeping me going, and a burning desire to still keep on to the very best of what is left in me.'

Then, later in the log: 'Be-bloody-calmed again! I've never known frustration like this voyage in my life before. What a complete waste of time and effort . . . there must be a dozen or so yachts ahead of me.'

On 27 June came the most poignant entry: 'Have just heard that Williams has crossed the finishing line. Good luck to him. Will press on as hard – coming second or third is better than a jab in the eye with a sharp stick.'

The handwriting was as firm, well-formed and neat as it had been throughout.

'I'm going to enjoy that champagne once the pick is down and settled.' The final entry before he sailed into Newport.

But there was one other thing he had written, earlier in

his book, which was to show perhaps more than anything what the voyage had done to him.

'I hope that one day, partly as the result of experiences such as these, I might be called a gentle man.'

And I cannot help thinking of him, minutes after his arrival, grabbing my arm as we stood in the well of his boat, when he whispered very quietly: 'You know, that was the one I wanted to win. God, I tried man. I really tried.'

In the days after his arrival, Dalling simply would not be drawn on the question of Williams sailing north of the Nantucket Light. He had dismissed as negligible the advantage the Englishman had gained by being computer-aided, and he was willing to say that Williams had the better boat – there was no false modesty about this, and in any case he must have known that with at least a fair wind over the last four days he could have won. But he refused to discuss the Light.

It could be that he genuinely felt that Williams had gained no particular advantage. Equally, it could be that he suspected the advantage to be as great as forty-eight hours. This figure has been suggested to me by two other competitors who arrived in Newport, and who said that if they had been close enough behind, as Dalling was, they certainly would have protested.

They base their estimate of time saved not simply on the fact that Williams was two or three hours' sailing time north of the Nantucket Light, which is all he was in fact, but on the navigational plan he was able to follow sailing down the coast of New England, closer inshore than he could possibly have been had he intended to sail out and around the Light. Inshore, he was likely to find the more favourable currents and winds.

Clearly it is not in Dalling's nature to protest, and he was greatly admired in Newport for not doing so, despite the many pressures brought to bear on him by misguided well-

Silvia II *was dismasted after only three days. André Foezon returned to Plymouth, set off a second time on June 12th, and took 29 days to cross to Newport.*

Jean-Yves Terlain sails his English-designed glass-fibre yacht, Maguelonne, of the Elizabethan 35 class. It was the first French-built boat to finish the Race.

wishers and backers. But there was no doubt in his mind that he could have won the race. One crucial error of judgment off Newfoundland, or perhaps it was just another piece of unavoidable bad luck, cost him that victory he wanted so much.

When he was off Cape Race and he and *Lipton* (I estimated from comparing their positions afterwards) were probably less than a hundred miles apart, Dalling elected, instead of sailing south-west like Williams, to sail south-east in search of stronger winds. There were no winds, and he was becalmed for two days. As he said: 'That's how it goes.' A reporter added: 'Without a computer.'

For the first two weeks of the race, *Cheers*, with Follett pursuing his southerly trade winds course, had gone like a speed boat, and had the wind held Follett must have had a very fair chance of hitting his twenty-one day target. But he also lost the breeze, and was becalmed.

On her best day, *Cheers* sailed over two hundred and fifty miles. *Lipton*'s best was less than two-twenty, and *Voortrekker*'s about two-twenty-five. Follett, on a southerly route that has always been thought to be the longest way round, sailed fewer miles through the water than either of the other boats. Three thousand four hundred to Williams's three thousand nine. A mark to Newick, who always contended that in good winds (when *Cheers* did not have to 'tack') she would prove a formidable sailing boat.

With the first three boats home, and little more news of the next, Gavin Young who is also a correspondent of *The Observer*, came up from New York to help report the intriguing stories of the three very different sailors. Young is not a yachtsman, and saw them as brave men who had just performed a remarkable feat. I know that he as much as I was impressed above all with their profound respect for the sea.

8
The long wait

Newport is a sort of rich man's Cowes without the brass buttons. The city is surrounded on three sides by the sea, has a large naval base and is very much New England. Inevitably, everything is done on a grand scale, and when a Rhode Island hostess invites you to drinks to meet a few people, you can be sure that there will be at least five hundred guests. If the 'martini hour' is something typically American, then Newport is undoubtedly a most American city.

The International Set, fortunately, have not moved into Newport – the weather, being almost worse than England's, hardly encourages them. And yet it is a remarkably international place. Half Australia seems to have stayed behind after the two Antipodean America's Cup challenges, and in West Pelham Street you hear more Australian accents than you do American. Emigrants from this country are scarcely less numerous, and find that the tip of Rhode Island is almost home from home.

Robin Wallace, an English doctor who trained in Boston and now has a practice in Newport told me 'In some ways it is very much like England here, and there are no pressures on you to become Americanised.' He backs Britain, drives an M.G. and refuses to allow even a trace of the curiously nasal New England accent to creep into his speech. Wallace is a keen yachtsman and gave me considerable assistance in the Press office at the Port o'Call Marina: he probably did as much as anyone to see that information about the competitors was disseminated as widely as possible.

Dunning and his colleagues at the Port o'Call deserve medals for their sterling work. Dunning is also an expatriate

Englishman, and he, more than any other single person, gave Smith and me enormous help. He stayed up all night on occasions to ensure that arriving competitors were properly received at his marina. When there was no room for the yachtsmen or their friends and relations at his inn, then he personally arranged accommodation for them elsewhere.

Murphy (the Ida Lewis Secretary) and some of the members of the club put themselves to considerable inconvenience to house, counsel and advise yachtsmen and journalists. Murphy and Dalling later sailed *Voortrekker* to New York, from where the yacht was shipped back to South Africa.

But before the competitors started to leave Newport there was a tumultous round of receptions and parties, and those of us who were waiting for the smaller boats to arrive were caught up in the celebrations. The hospitality of the people of Newport was overwhelming, and each competitor was moved by it.

The fourth boat to sail in was *Spirit of Cutty Sark*. Leslie Williams arrived on a brilliantly sunny Sunday, as he had predicted, exactly two weeks after he had last spoken to me by radio.

'My main fear,' he told me, 'was not the storm, but the ice off Newfoundland. That ice really scared me, and all my basic instincts as a seaman rebelled. I just did not want to go on. The ice reports told me where to look but the fear was always there.

'One day off Cape Race, although I was utterly composed, I just hove to for half an hour. I don't know why, but I sensed something. After thirty minutes or so I was all right again and sailed on. A premonition of disaster, perhaps, or some sixth sense.'

Cutty Sark was the first production boat to finish, and Williams was immensely loyal to her. The expert opinion after the race was that in a one-off yacht like *Lipton* he could

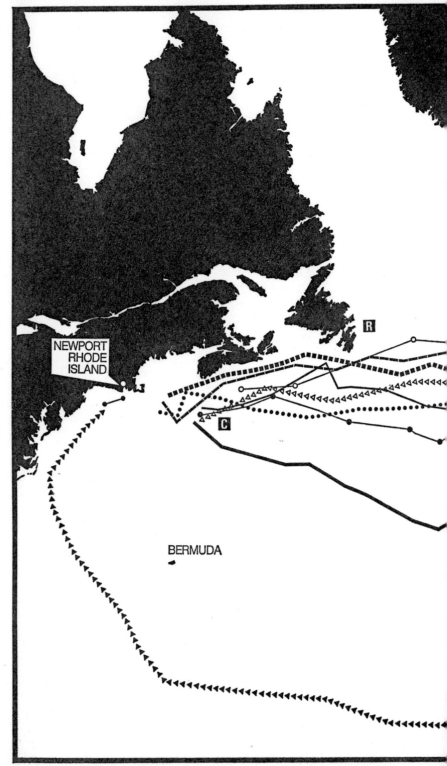

Fig. VIII The Routes taken by some of the Competitors.

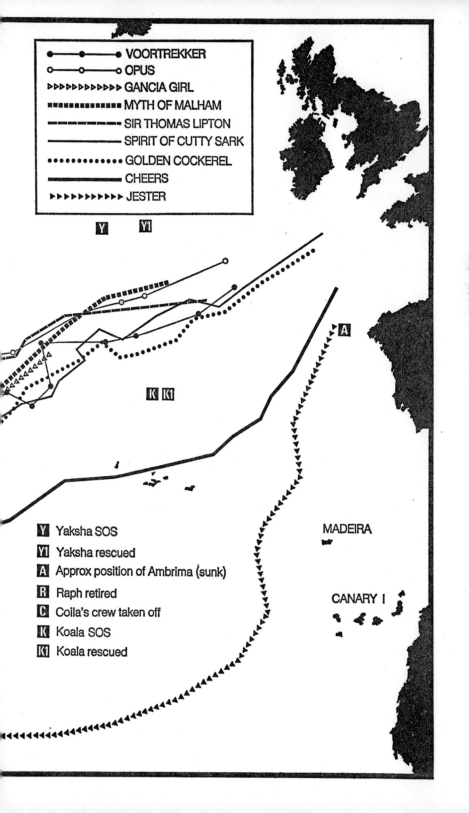

	VOORTREKKER
	OPUS
▷▷▷▷▷▷▷▷▷▷▷▷	GANCIA GIRL
▬▬▬▬▬▬▬▬▬▬	MYTH OF MALHAM
─ ─ ─ ─ ─ ─	SIR THOMAS LIPTON
────────	SPIRIT OF CUTTY SARK
••••••••••••	GOLDEN COCKEREL
━━━━━━━━	CHEERS
▶▶▶▶▶▶▶▶▶▶▶	JESTER

Y Yaksha SOS

Y1 Yaksha rescued

A Approx position of Ambrima (sunk)

R Raph retired

C Coila's crew taken off

K Koala SOS

K1 Koala rescued

MADEIRA

CANARY I

have won quite easily. He defended *Cutty Sark* staunchly, saying that he had chosen to sail her and that she had served him well. Of that there could be no doubt, for even with a totally unsuitable rig – a single mast with one gigantic foresail on such a heavy boat cannot possibly permit one man to do full justice either to himself or to the yacht – she covered the course in less time than Chichester had done in 1964, and in only two more days than Tabarly. With a ketch rig she would surely have beaten Tabarly's 1964 time.

Spirit of Cutty Sark was a beautiful boat to sail and even in rough conditions (because of her heavy displacement) did not beat down on the waves or give the impression that she was about to shake herself to pieces – an impression which *Voortrekker* often gave – and consequently she offered her passengers a most comfortable sail.

The *Cutty Sark* team were extremely popular in Newport and they painted the town fairly red. On 4 July, Williams declared independence and fired every flare on board. For some hours the waterfront was wreathed in the not very delicately perfumed pink smoke of the Very lights.

Two days after Williams's arrival, Howell radioed that he was about to cross the finishing line in *Golden Cockerel*. The catamaran eluded searchers in the early evening mist, but the Australian accent and graphic language more than adequately described the course he was taking.

When the boat in which I was searching had made visual contact, Howell grabbed a megaphone and shouted 'That was a bloody awful trip. I lost a stone and a half on this bloody boat.'

Ashore, I asked him what he had done in the great storm. 'I just drank booze.' He drank twenty-three dozen cans of beer in thirty days.

But he made some serious comments about *Golden Cockerel*, and catamarans in general. 'They are very exciting boats, and they're fast with two men on board. With two on board in this race I could have won it. As it was, I spent all my time reining her in.

'I don't think an ocean racing cat such as I've got here is a singlehanded boat. She just goes too fast and gets out of hand. She runs away with one man all the time and you have to rein her in. It was an effort to get the boat into the wind at all without rolling her.'

Howell had his experiences in 1964 to guide him when he decided on his route (Fig. VIII). 'I had planned to head for the tail of the Grand Banks and then skirt the northern edge of the Gulf Stream until I could head directly for the Nantucket Light Vessel. I tried to keep rigidly to this plan which was dictated by my experience and Tabarly's in 1964.

'However, conditions did not repeat themselves and instead of the light reaching winds that I had hoped for in the last thousand miles I had to bash into strong headwinds. I was finally dished by a completely wrong U.S. Coastguard weather forecast on 29 June, which trapped me in gale force winds the following day on the southern edge of George's Bank. This cost me two days.

'It was a rough race, much harder than the 1964 one, and the seas in the storm of 11 June were the worst I have been through in a yacht.

'I have no excuses. Although I only had thirty hours of free wind during the whole race, I realise that on such a course as I chose this is a possibility that must be faced. However, the catamaran stood up to the bashing amazingly well. I only lost one tier and one shackle on the voyage. A butt joint where a frame joined the keel sprang when I was hurled bodily to leeward during a gale on the George's Bank – this was the builder's fault as it should have been scarphed in.

'Much more development work on self-steering for high speed offshore racing multihulls is needed before they can really be successful in singlehanded racing. I found myself continually changing sails to meet the changing weather conditions. I changed sails two hundred and eighty-one times during the race.'

In spite of this, he still bettered his 1964 time by nearly a week, and seemed in very good spirits. Would he, as had been rumoured, sail *Golden Cockerel* round the world? 'Not bloody likely. At least not alone.'

The Australian dentist was immensely popular in Newport, and was remembered by the locals. He is a jovial man, ever ready with a funny, probably dubious, story. But he is also a very experienced sailor who knows a great deal about yachts and the sea. He is patient in explaining technicalities, and never forces his opinion, though he possibly knows more about the Atlantic than any other competitor except Follett. A boisterous man, with a huge grin and a typically Australian disregard for convention, he is above all, like Follett, a humble man with a profound respect for the sea.

He and Leslie Williams turned their boats into floating hotels, and their corner of the marina became the focal point of social life for the next week or so.

One of the most impressive performances of the race was Cooke's. He sailed *Opus* into Newport just thirty-four days after leaving Plymouth, a remarkable passage by any standards. People on the waterfront stared in disbelief at the dumpy little sloop. 'She is very seaworthy,' said Cooke.

He was obviously moved by the welcome he received. 'I never expected anything like this.'

'The worst moment of the trip was when our forestay snapped.' (He used the royal plural whenever he talked about the voyage – 'I could not have got here without Opus, and she needed me'.)

'The forestay was sheered right off. I could not go up the mast for two days because of the seas and I had to set my foresail flying. The halyard and the winch broke as well.

'The winds were the strongest I have ever experienced but then at the end of the race I encountered calm after calm. I saw a gigantic iceberg and must have sailed right through the ice – but that was the shortest route here.'

Five hours after Cooke's arrival, Minter-Kemp sailed *Gancia Girl* across the finishing line – the first trimaran by a long way. The Army captain looked remarkably fit and unruffled, and seemed the least exhausted of any of the yachtsmen who had finished.

'The rough weather was no problem, I just went on cooking *vol-au-vent*,' he joked. 'And in the storm I took everything down and went to bed.

'Certainly I met some big seas, but she's a super boat and she's got no vices at all. My major setback was the loss of my self-steering gear, it broke first day out of Plymouth.'

He had hoped for a fast sail in from Sable Island. 'When I heard on the radio of *Lipton*'s arrival in Newport I had only six hundred and fifty nautical miles to go and a fair wind was giving me one hundred and eighty a day. I decided on a thirty-day trip. A fatal thing to do – the wind died to nothing and worse still the sea turned bright blue and was strewn with weed. I was in the Gulf Stream and probably being carried east. All I could do was curse the current chart for June, which put the Stream much farther south, and pray for wind.

'I got it, fifty knots of it. "This is not right" I said to myself, and in retaliation the wind died for two more days. Eventually a faint breeze carried us back into the Labrador Current, and we were being set down towards the Nantucket Light.'

Myth of Malham was the eighth boat to finish, and Bevan complained bitterly of the calms that had delayed him virtually all the way in from Cape Cod. 'Becalmed. Becalmed. Wind gone. Ditto.' These were almost the only entries in his log for six days at the beginning of July. He had been very frustrated by the weather and was disappointed with his slow time. On being congratulated, he replied 'What for?' His time of thirty-six days was, in view of the windless conditions, a very fine one.

De Castelbajac arrived on 8 July after consuming seventy bottles of wine during the crossing. He had trouble in navigating *Maxine*, as both his watch and his chronometer broke and his radio receiver was not working properly.

His countryman Terlain finished next and said that he was very pleased with his French self-steering gear which worked so efficiently that he was able to sleep for almost eight hours a day. The tail end of the June storm had knocked *Maguelonne* down, and all the electrical and radio equipment had been broken, but his sloop behaved perfectly in the heavy seas.

Having spent the last four years on duty in Singapore, Burgess thought that he would not much like the cold of the North Atlantic, and he chose a southerly route. The self-steering gear on *Dogwatch* proved inadequate and he had to steer most of the way with his hand on the tiller. 'There was one storm,' he said, 'when the winds rose to force 11 or 12.

'Then there were exasperating calms during which I shouted and swore at the boat and the weather until the paint peeled.'

One of the more unusual arrivals in Newport was by aeroplane, not by boat. Gliksman left *Raph* in San Pierre, where her rudder was being repaired, and flew down to Rhode Island. He wanted to write the stories of some of the competitors for his magazine, to take pictures and to visit the Jazz Festival.

Foezon missed the Festival, but arrived in the very fast time of twenty-nine days. *Silvia II* in fact came twelfth on elapsed time (forty days); for she had been dismasted on 4 June, had returned to Plymouth and had not set out again until eight days later. Her corrected time on the second attempt was twenty-one days – a day less than *Voortrekker*'s corrected time.

The Frenchman had taken the Northern Route and experienced bitterly cold weather with temperatures of 5 °C. On 21 June his challenge in the race nearly came to an untimely end, and the memory will haunt him, he said, for the rest of his life.

While changing jibs on the foredeck, he slipped, lost his grip and grabbed at the jib halyard. But it was secured only by a piece of cord, and he fell into the sea. Luckily, the jib was trailing in the water, and Foezon – who refuses to wear a safety line – was able to haul himself back on board.

The first of the tiny Swedish boats, *Fione*, slipped unnoticed into Narraganset Bay late in the evening of 11 July. Enbom tied up to some pilings so that he could extend one of the thirty minute cat naps with which he had had to make do during the voyage, and he slept right through the night. On Friday morning he walked nonchalantly round to the Port o'Call to sign in.

Enbom was clearly delighted with the performance of his boat and said that she had made light of the heavy weather during the voyage. He hoped that there would always be 'little' entries like his, and that sponsorship would not kill the spirit of the race.

The German architect Hehner declared that he was glad to be in Newport after a frustrating forty-two days at sea. 'For six of them I was totally becalmed without a drop of wind, and I spent the time re-stitching *Mex*'s sails many of which had been torn.'

Pakenham took the Azores route, and was becalmed. He finished fifteenth in *Rob Roy*, and he thought that he had failed in his endeavour. A roar of dissent from the welcoming crowd assured him that he was not a failure, and he was invited to take a local church service on Sunday.

'After a while you run out of booze and things to read,'

was Forbes's weary comment as he stepped ashore from *Startled Faun.* 'This is not a race for big multihulls. You're fighting against headwinds all the way.'

The other trimaran, *Amistad*, was also troubled by headwinds. Heavy fog near Newport almost caused Rodriquez's yacht to run aground. 'I was cooking some stew when all of a sudden I heard surf on three sides. I had come into an inlet. So I threw over my anchor and waited for the fog to clear. It's lucky the hull wasn't punctured.'

During the long voyage he passed the time 'doodling sketches of the larger trimaran I want to build next year'. In spite of the headwinds he encountered in the race, he said he would continue to sail multihulls as they are potentially faster.

Goodwin II arrived after fifty days, but Mattsson was disqualified for taking on supplies during the race. He had had fresh water dropped to him, and Race Rule 21 states that no stores may be received during the race.

Willis had been doing extremely well when he was taken ill at the end of June. His drinking water had become contaminated and his condition was so serious when he radioed for assistance that two para-medics were dropped to *Coila* to assist. He was later taken off the trimaran and she was towed into harbour.

There was still one boat at sea which had not been sighted since she left Plymouth. With all the other yachts safely in port, concern for Richey's safety grew daily. By the end of his eighth week at sea panic and alarm in the Press took over from common sense, and it was said that he was 'overdue.' 'Over what?' said Richey when he heard of the fuss, on his arrival after fifty-seven days.

Richey is executive secretary of the Institute of Navigation in London, and did not enter the race with the intention of winning. His object in taking part was to try and find a more economical passage across the Atlantic.

FINAL PLACING IN THE 1968 RACE

Yacht	Crew	Elapsed Time			Handicap			Place
		D.	H.	M.	D.	H.	M.	
Sir Thomas Lipton	G. Williams	25	20	33	22	16	51	1
Voortrekker	B. Dalling	26	13	42				*
Cheers	T. Follett	27	00	13	25	02	44	4
S. of Cutty Sark	L. Williams	29	10	17				**
Golden Cockerel	B. Howell	31	16	24	24	16	14	3
Opus	B. Cooke	34	08	23				*
Gancia Girl	M. Minter-Kemp	34	13	15	28	11	01	8
Myth of Malham	N. Bevan	36	01	41	27	08	36	6
Maxine	B. de Castelbajac	37	13	47	27	18	17	7
Maguelonne	J.-Y. Terlain	38	09	10	27	06	20	5
Dogwatch	N. Burgess	38	12	13	29	12	29	9
Silvia II	A. Foezon	40	00	16	24	14	13	2
Fione	B. Enbom	40	14	13	32	00	04	11
Mex	C. Hehner	41	10	46	30	10	20	10
Rob Roy	S. Pakenham	42	03	49				*
Startled Faun	C. Forbes	45	10	08				*
Amistad	B. Rodriquez	47	18	05				
Goodwin II	A. Mattsson	50	19	48				
Jester	M. Richey	57	10	40	37	19	43	12

* Multihulled boats were not eligible for the Handicap Trophy.

Goodwin II finished 18th but was not placed in the race, she was disqualified under Rule 21.

Silvia II was dismasted and had to return to Plymouth, set out again on 21 June and her actual sailing time for the crossing was 29 days.

Sir Thomas Lipton did not have a handicap as she was not presented for scrutineering before the race.

He took an extreme southerly course, and for the first time in this race a yacht attempted to sail a genuine Trade Wind Route. He said ' . . . that *Jester* arrived last in the race may reflect on the choice of route – though I would attempt a similar trajectory in a boat of competitive size. Her performance in the widely differing conditions she encountered could not have been bettered. She was delayed by prolonged calm.'

'It is unfortunate that her prolonged crossing this time led to anxiety and plans for a search. A transmitter would indeed have allayed the fears. On the other hand it would not remotely have added to the boat's safety. . . . Survival is not the principal difficulty of such a voyage.'

So, in the end, all thirty-five of the yachts were accounted for. Where the experts had prepared themselves for news of at least one fatality, there was none. The yachtsmen had done something for no other reason than that they wanted to. Some of them turned their minds immediately to plans for the Round Britain Race and the next Transatlantic Race.

For the moment, though, they were satisfied, their achievement was considerable and complete. The Atlantic venture was over.

9
Conclusion

It has been said that this year's race was the last of the great adventures, and that there will never be another like it. Certainly the third race had drama, excitement, tension, humour, emotion and very real human interest. All these will be present in 1972, and whenever men take on single-handed the danger and treachery of the Atlantic.

What the purists fear is that it now having been proved that sponsors are willing to invest vast sums of money in backing an entrant, the man who wants to stimulate, like Hasler, the design of small sea-going sailing boats will be frightened out of the hunt by the big chaps with the thick cheque books. And what then of the future of the event?

The Race Committee, and *The Observer*, are keenly aware of the dangers of over-sponsorship. But just as they would not exclude a small boat and its skipper because they were unsponsored, how could they bar a boat whose owner had the skill, and business acumen, to persuade sponsors to give him thirty-thousand pounds.

As the Rules and Conditions of Entry state, the object of the race is to encourage the development of suitable boats and gear, supplies and technique for single-handed ocean crossings under sail. The race is intended to be a sporting event. In discussing the problems raised by this last race, the Committee has been advised by Odling-Smee that they must insist that the race should remain a sporting event, and that they should reserve the right to refuse an entry if it appears that the primary object of that entry is to promote a commercial project not connected with the objects of the race.

This may sound somewhat arbitrary, but then it could be

argued that the decisions of almost all committees are arbitrary, and who else is to determine the eligibility of an entry if not the Race Committee. Nor would it be possible to impose any financial ceiling on the sum of money allowed to be invested in a given challenger. Neither would it be fair to exclude commercial organisations from the class of prospective sponsors. As Odling-Smee states in the draft Rules for 1972, '. . . I think we must trust to the good taste of the Sponsors . . .', and he and his Committee still reserve the right to be the arbiters of that taste.

There were many complaints about the 1968 race, which were communicated in the most part to *The Observer*, to the newspaper's Press officers, to the Royal Western Y.C., and not a few to other newspapers. Basically they were all variations on a single theme: safety.

At the height of the 'safety controversy', Blondie Hasler wrote in *The Observer* what to my mind was the most brilliant rejoinder to any critic of the race. Parts of it are here reproduced at some length.

'Criticism of the event seems to revolve around a few general themes:

    1. Multihullers should be excluded.
    2. The Committee should be more strict in barring boats, the design of which they consider to be unseaworthy.
    3. The Committee should be more strict in barring skippers who are inexperienced, or female.
    4. There should not be such a race, anyway.

'Various taxpayers have also protested about the cost of air/sea rescue operations, but it seems possible that what they really object to is rescuing *foreigners*. They don't seem to have got in the same tizzy over the equally expensive air/sea

*Geoffrey Williams (left), Bruce Dalling (right)
and the author (centre) sail* Sir Thomas Lipton *off
Newport, Rhode Island. Dalling thought Williams'
boat much faster than his own* Voortrekker.

*Tom Follett (left), Geoffrey Williams (centre)
and Bruce Dalling (right). Williams and Dalling
were already in Newport and waiting to
greet Tom Follett when he sailed* Cheers *in.*

search and escort operations mounted on behalf of Sir Francis Chichester or Sir Alec Rose, both of whom were engaged in recreational voyages under sail. Would it be outrageous to suggest that rescuing foreigners from the sea does more to further international goodwill than our traditional occupation of trying to drown them in it?

'The rules and conditions of entry of the race are not immutable, but get reviewed by the organisers after each race, and tend to get stricter each time.

'It seems possible that further restrictive conditions will be imposed for any future race, and although I am not on the Committee I would like to say something about the background against which these difficult decisions have to be made.

'Since 1960, the race has grown rapidly in size and importance, with entries from ten different countries this year. This international interest sets up strong pressures on the organisers, who have a pretty rough ride in the three months before the start. What, for example, will be the French reaction if a fast French trimaran is barred from the race because the British Committee don't think she looks quite right for the job? Perfidious Albion reacting to the French victory of 1964?

'Again, whenever any entrant feels the risk of rejection, he invariably announces that he will sail over the course anyway, and there is no legal means of stopping him. Faced with this sort of determination, it would not be surprising if the Committee were sometimes to feel that it is easier to admit a boat as an official entrant than to risk having him turn up at the finish in record time after having been rejected.

'There is also a human side to it: the Committee are themselves amateur yachtsmen, and it is appropriate for sportsmen to make sporting decisions. They cannot, and should not, fail to consider the feelings of the fanatically keen, often impoverished, enthusiast who has sometimes staked a couple of years of effort, and all his money, in an attempt to sail this race.

'This race gives a powerful shove to the process of development by admitting yachts of all types, and then providing a hard testing ground in which the more extravagant claims of some designers, and some salesmen, are neatly cut down to size by the sea itself.

'Should the Committee have barred those skippers who have since had to be rescued? Certainly not Joan de Kat, who is said to have made a previous singlehanded Atlantic crossing. Fraulein Baumann may have been a marginal case, but the Committee tested her navigational ability, and she does not in fact seem to have made any particular mistakes in handling the boat.

'My personal view is that the race is being guided along the right lines, and that it should be allowed to continue and develop. No doubt it will always stir up its share of criticism, but so do many other progressive activities. What heartens me is the large number of people, both seamen and landsmen, who think that it is a worthwhile thing.'

Hasler's theme is that you cannot have success without failure. If men are to be allowed to try and prove something, then it is inevitable that occasionally they will fail.

In spite of that, it is a fact that many people do consider the race to be a worthwhile venture. For this reason if for no other, the future of the race is assured. For the next race, it is almost certain that bigger and better boats will be entered, with more sophisticated equipment on board. As Geoffrey Williams told me, his new boat is already on the drawing board. But there will always be the Tom Folletts and the Brian Cookes about, and there is a good chance that one day one of them will win.

There may still be people who maintain that this race serves no useful purpose. I have tried to show how in a practical way it contributes more than merely to the refinement of yacht design and gear. If a yacht like *Sir Thomas Lipton* can make an extremely fast, seaworthy passage across the Atlantic with only one man on board, then what of its speed, safety and value with a full crew? What of its uses as a vessel for sail training?

If *Spirit of Cutty Sark* could be sailed through a gale by a lone yachtsman, and arrive safely at the other side, what of her safety for pleasure cruising, and, as a standard design with normal fitting, what of her uses as a family boat?

And if all that sounds too commercial, then why is not the race worth competing in solely for the sake of taking part. It is a form of endeavour no less demanding or worthwhile or exciting or dangerous than mountaineering – though there are some people who would argue that to climb a mountain serves no useful purpose. Nor then, I would have to claim, does anything more strenuous than a social set of mixed doubles.

The astonishing thing about this race, and the challenge of the Atlantic, is the effect it has on those who answer the challenge. A mountain climber conquers his mountain – he fights it, he struggles with it, he is cowed by it, but in the end he vanquishes it, and is pictured, arms akimbo or flag aloft, bestriding it triumphantly – the victor.

The men who sailed into Newport last summer gave me no such impression. There was a victor, but he was victor in the race, and had beaten only his fellow competitors. There was no talk of having conquered the Atlantic. For the sea is not something you master. The lone yachtsmen gave only the impression of having arrived in a spirit of the most profound humility. There was personal satisfaction, certainly. The satisfaction of a job well done, of an ambition realised. But there were no flags aloft.

The nineteen finishers in the race had fought with the Atlantic, struggled with it, been cowed by it, and in the end had worked in concert with it, never trying to assert themselves. I could see why Follett was such a humble man when I had first met him in Plymouth – he knew the Atlantic already. And I felt that the winner had become a much more humble man, for having learnt to live with and, even if he would not admit it, love the sea.

That is what the thirty-four men and one woman were about in Plymouth Sound last first of June.

APPENDIX 1

Scale drawings of yachts in the 1968 Race

($\frac{1}{13}$ inch = 1 foot)

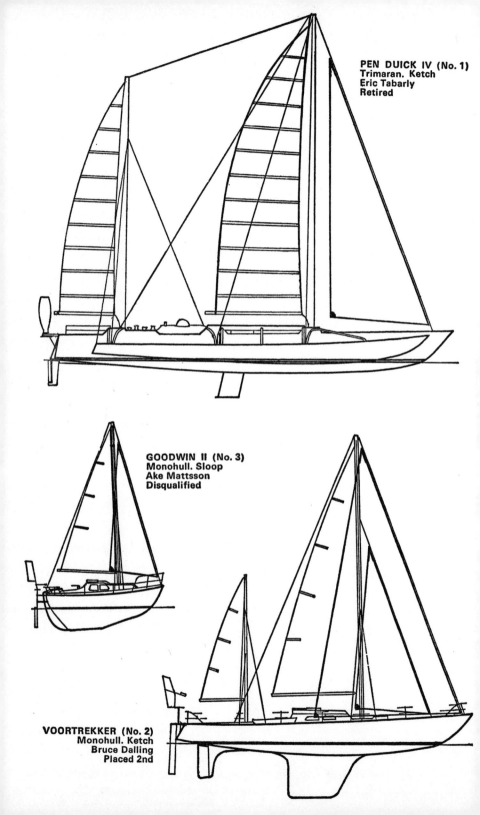

PEN DUICK IV (No. 1)
Trimaran. Ketch
Eric Tabarly
Retired

GOODWIN II (No. 3)
Monohull. Sloop
Ake Mattsson
Disqualified

VOORTREKKER (No. 2)
Monohull. Ketch
Bruce Dalling
Placed 2nd

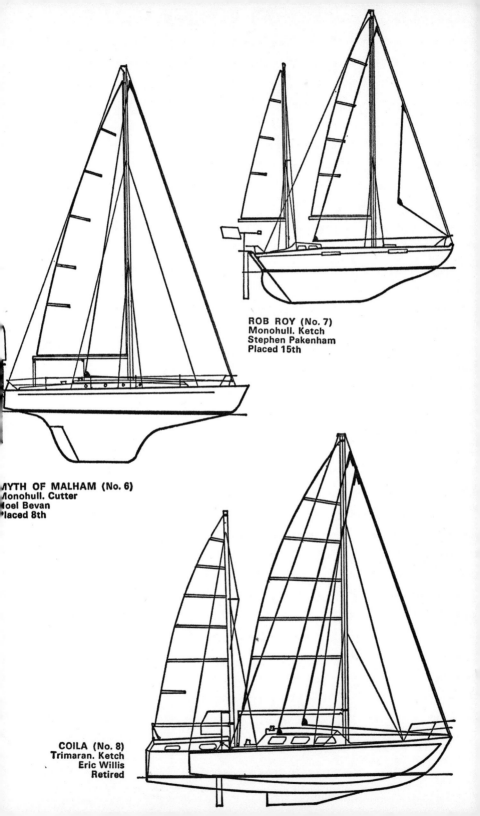

ROB ROY (No. 7)
Monohull. Ketch
Stephen Pakenham
Placed 15th

MYTH OF MALHAM (No. 6)
Monohull. Cutter
Noel Bevan
Placed 8th

COILA (No. 8)
Trimaran. Ketch
Eric Willis
Retired

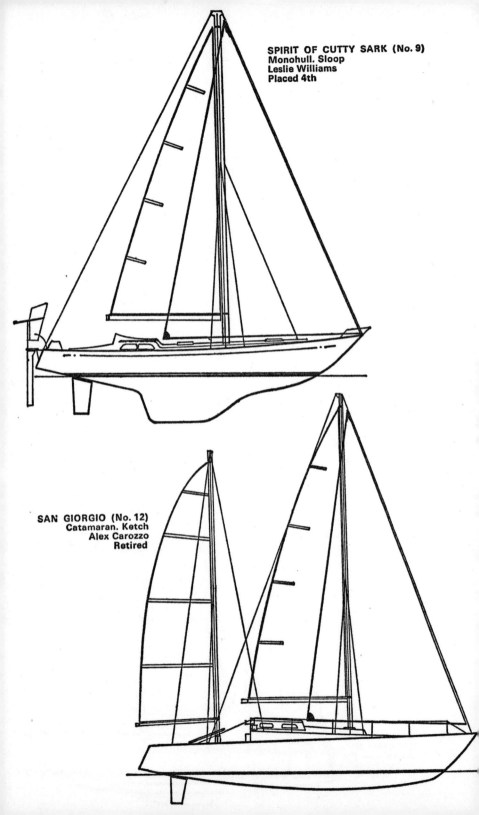

SPIRIT OF CUTTY SARK (No. 9)
Monohull. Sloop
Leslie Williams
Placed 4th

SAN GIORGIO (No. 12)
Catamaran. Ketch
Alex Carozzo
Retired

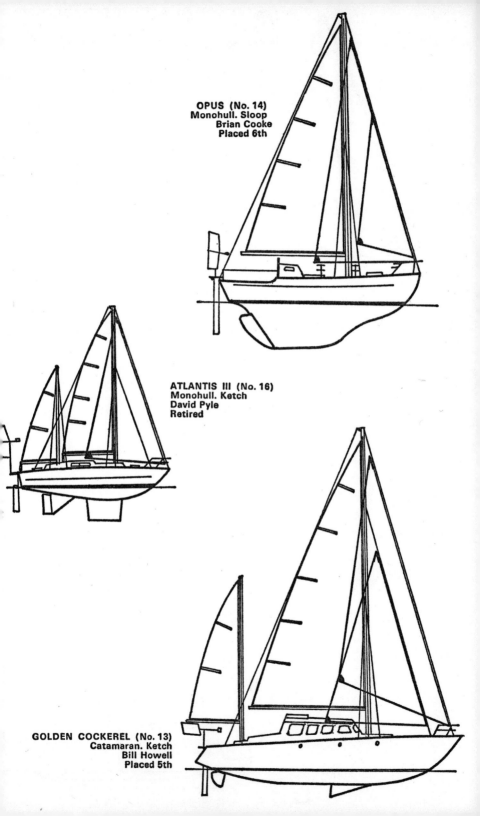

OPUS (No. 14)
Monohull. Sloop
Brian Cooke
Placed 6th

ATLANTIS III (No. 16)
Monohull. Ketch
David Pyle
Retired

GOLDEN COCKEREL (No. 13)
Catamaran. Ketch
Bill Howell
Placed 5th

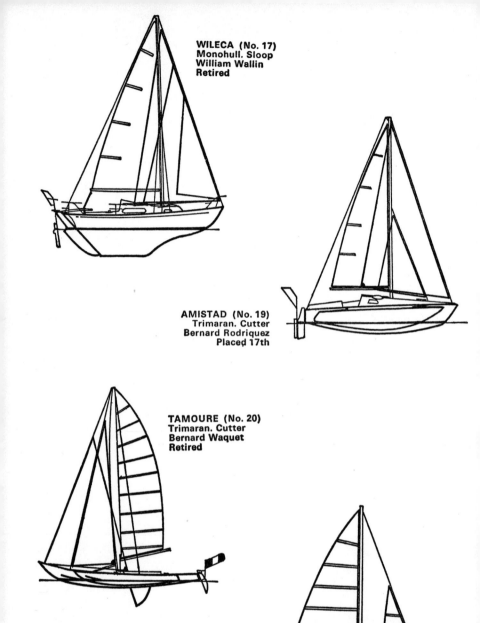

WILECA (No. 17)
Monohull. Sloop
William Wallin
Retired

AMISTAD (No. 19)
Trimaran. Cutter
Bernard Rodriquez
Placed 17th

TAMOURE (No. 20)
Trimaran. Cutter
Bernard Waquet
Retired

KOALA III (No. 21)
Trimaran. Sloop
Edith Baumann
Sank

FIONE (No. 24)
Monohull. Sloop
Bertil Enbom
Placed 13th

ZEEVALK (No. 23)
Monohull. Sloop
Robert Wingate
Retired

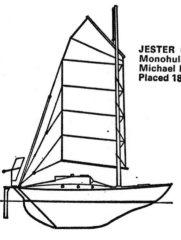

JESTER (No. 27)
Monohull. Chinese lug
Michael Richey
Placed 18th

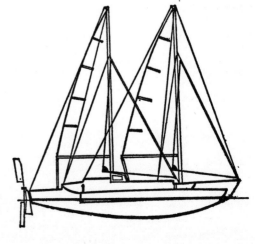

WHITE GHOST (No. 25)
Trimaran. Schooner
Michael Pulsford
Retired

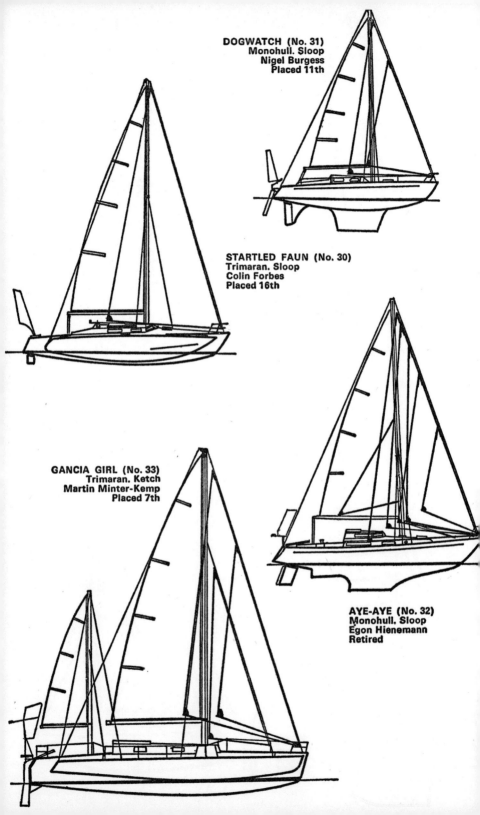

DOGWATCH (No. 31)
Monohull. Sloop
Nigel Burgess
Placed 11th

STARTLED FAUN (No. 30)
Trimaran. Sloop
Colin Forbes
Placed 16th

GANCIA GIRL (No. 33)
Trimaran. Ketch
Martin Minter-Kemp
Placed 7th

AYE-AYE (No. 32)
Monohull. Sloop
Egon Hienemann
Retired

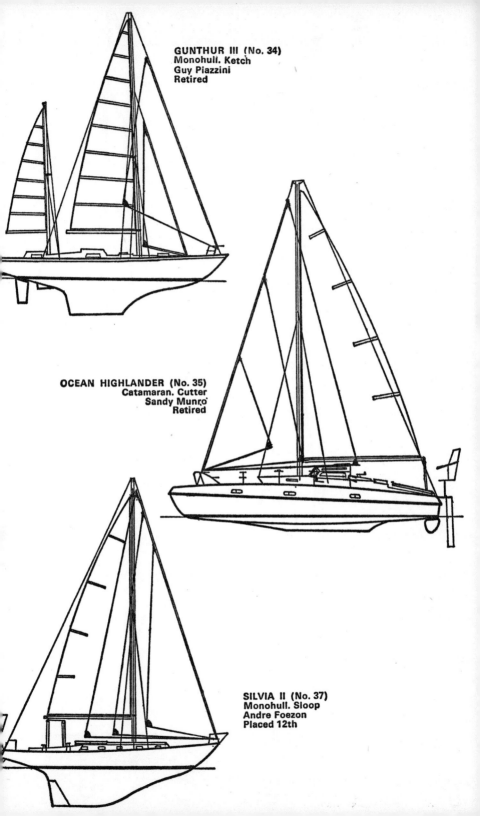

GUNTHUR III (No. 34)
Monohull. Ketch
Guy Piazzini
Retired

OCEAN HIGHLANDER (No. 35)
Catamaran. Cutter
Sandy Munro
Retired

SILVIA II (No. 37)
Monohull. Sloop
Andre Foezon
Placed 12th

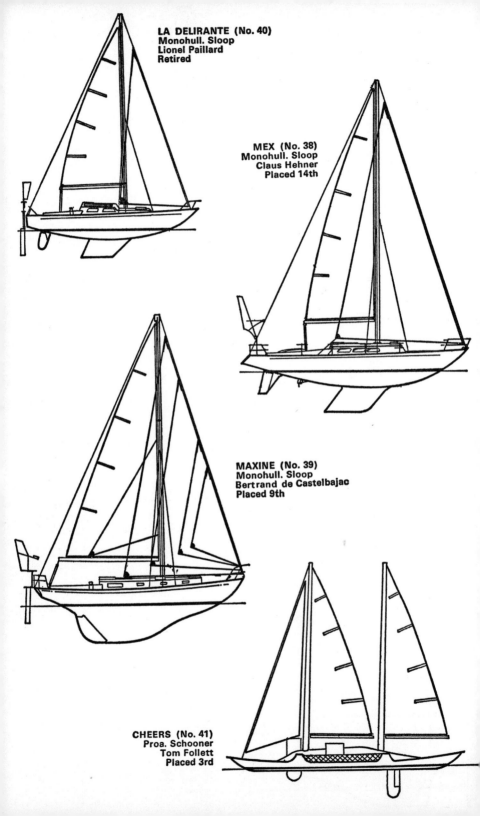

LA DELIRANTE (No. 40)
Monohull. Sloop
Lionel Paillard
Retired

MEX (No. 38)
Monohull. Sloop
Claus Hehner
Placed 14th

MAXINE (No. 39)
Monohull. Sloop
Bertrand de Castelbajac
Placed 9th

CHEERS (No. 41)
Proa. Schooner
Tom Follett
Placed 3rd

RAPH (No. 42)
Monohull. Ketch
Alain Gliksman
Retired

YAKSHA (No. 43)
Trimaran. Sloop
Joan de Kat
Sank

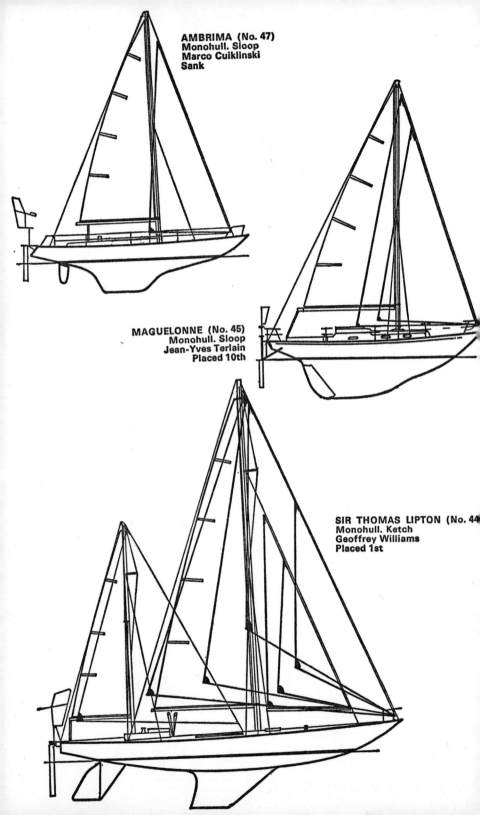

AMBRIMA (No. 47)
Monohull. Sloop
Marco Cuiklinski
Sank

MAGUELONNE (No. 45)
Monohull. Sloop
Jean-Yves Terlain
Placed 10th

SIR THOMAS LIPTON (No. 44
Monohull. Ketch
Geoffrey Williams
Placed 1st

APPENDIX 2

List of Entries for the 1968 Race

APPENDIX
List of Entries

No.	NAME OF YACHT	HELMSMAN	COUNTRY	DESIGNER	BUILDER	TYPE
1	PEN DUICK IV	Eric Tabarly	France	J. Rouillard	Chantiers et Ateliers de la Perriere, Lorient	Trimaran
2	VOORTREKKER	Bruce Dalling	South Africa	E. G. van de Stadt	Thesens, Knysna, Cape Province	Monohull
3	GOODWIN II	Ake Mattsson	Sweden	Lage Eklund	A. B. Fisksatra, Varv	Monohull
6	MYTH OF MALHAM	Noel Bevan	G.B.	Laurent Giles	H. McLean and Son, Gourock	Monohull
7	ROB ROY	Stephen Pakenham	G.B.	G. A. Feltham	G. A. Feltham	Monohull
8	COILA	Eric Willis	G.B.	A. Mylne	Bute Slip Dock Co., Isle of Bute	Trimaran
9	SPIRIT OF CUTTY SARK	Leslie Williams	G.B.	E. G. van de Stadt	Tyler Boat Co., Southern Ocean Supplies	Monohull
12	SAN GIORGIO	Alex Carozzo	Italy	Owner	Owner	Catamaran
13	GOLDEN COCKEREL	Bill Howell	Australia	Rudy Choy	Contour Craft, Great Yarmouth	Catamaran
14	OPUS	Brian Cooke	G.B.	Robert Clark and owner	R. F. Austin	Monohull

2
for the 1968 Race

RIG	L.O.A. (Feet)	L.W.L. (Feet)	BEAM (Feet)	DRAFT (Feet)	SAIL AREA (Sq. ft.)	Displacement (Tons)	RESULT
Ketch	67·00	59·70	35·00	3·90	1284	4·90	Steering trouble
Ketch	49·25	39·50	11·50	3·60	900	6·50	2nd
Sloop	19·70	16·40	7·30	3·60	217	1·58	Disqualified
Cutter	37·65	33·50	9·25	7·25	680	8·0	8th
Ketch	32·40	27·00	9·00	4·80	550	7·00	15th
Ketch	50·00	45·00	25·50	4·18	1300	6·00	Taken ill
Sloop	53·14	38·75	13·08	7·33	1360	15·00	4th
Ketch	54·00	46·00	23·00	2·00	1200	7·00	Rudder trouble
Ketch	43·00	36·00	17·00	1·83	950	3·50	5th
Sloop	32·00	24·50	9·40	4·80	600	5·50	6th

No.	NAME OF YACHT	HELMSMAN	COUNTRY	DESIGNER	BUILDER	TYPE
16	ATLANTIS III	David Pyle	G.B.	B. C. Pyle and D. Pyle	Owner	Monohull
17	WILECA	William Wallin	Sweden	E. G. van de Stadt	Tyler Boat Co., William King Ltd.	Monohull
19	AMISTAD	Bernard Rodriquez	U.S.A.	Arthur Piver	Rodriquez & Nixon	Trimaran
20	TAMOURE	Bernard Waquet	France	Owner	Herve, La Rochelle	Trimaran
21	KOALA III	Edith Baumann	West Germany	B. Waquet	Craff Benodet	Trimaran
23	ZEEVALK	Robert Wingate	G.B.	E. G. van de Stadt	E. G. van de Stadt	Monohull
24	FIONE	Bertil Enbom	Sweden	Lage Eklund	A. B. Fisksatra, Varv	Monohull
25	WHITE GHOST	Michael Pulsford	G.B.	Eric Manners	Owner	Trimaran
27	JESTER	Michael Richey	G.B.	H. G. Hasler	H. Feltham	Monohull
30	STARTLED FAUN	Colin Forbes	G.B.	Arthur Piver	Contour Craft, Great Yarmouth	Trimaran
31	DOGWATCH	Nigel Burgess	G.B.	U. van Essen	Shipyard Visser, Loosdrecht	Monohull
32	AYE-AYE	Egon Hiene-mann	West Germany	Willy Asmus	Willy Asmus	Monohull
33	GANCIA GIRL	Martin Minter-Kemp	G.B.	Derek Kelsall	Multihull Construc-tion Co. Ltd.	Trimaran
34	GUNTHER III	Guy Piazzini	Switzer-land	A. Brenneur	Sampson, Sartrou-ville	Monohull

RIG	L.O.A. (Feet)	L.W.L. (Feet)	BEAM (Feet)	DRAFT (Feet)	SAIL AREA (Sq. ft.)	Displacement (Tons)	RESULT
Ketch	25·50	22·00	8·00	5·00	500	3·50	Steering trouble
Sloop	26·42	20·67	7·33	4·00	311	2·42	Sought warmer weather
Cutter	25·00	23·50	15·00	1·50	330	0·87	17th
Cutter	37·00	36·00	20·00	3·00	267	0·58	Unable to navigate
Sloop	39·40	37·00	27·00	2·00	534	2·94	Sank
Sloop	39·42	34·75	8·92	7·00	618	5·50	Leak in hull
Sloop	19·70	16·40	7·30	3·60	217	1·58	13th
Schooner	30·00	27·60	17·50	1·70	320	1·7	Rudder trouble
Chinese lug	25·75	19·50	7·16	4·00	246	2·5	18th
Sloop	33·00	31·00	19·25	2·60	487	2·75	16th
Sloop	27·80	24·00	7·60	4·90	240	3·00	11th
Sloop	33·30	28·10	9·66	4·40	551	4·7	Steering trouble
Ketch	42·00	39·00	22·30	4·30	600	3·00	7th
Ketch	40·90	30·80	11·00	5·50	600	6·3	Mast came loose

No.	NAME OF YACHT	HELMSMAN	COUNTRY	DESIGNER	BUILDER	TYPE
35	OCEAN HIGHLANDER	Sandy Munro	G.B.	G. Prout & Sons	G. Prout & Sons, Canvey Island	Catamaran
37	SILVIA II	Andre Foezon	France	Sparkman and Stephens	Chantiers Ziegler	Monohull
38	MEX	Claus Hehner	West Germany	Dick Carter	Abeking and Rasmusson	Monohull
39	MAXINE	Bertrand de Castelbajac	France	Alan Buchanen	R. J. Prior	Monohull
40	LA DELIRANTE	Lionel Paillard	France	Mauric and Gaubert	Qure	Monohull
41	CHEERS	Tom Follett	U.S.A.	Dick Newick	Dick Newick, St Croix	Proa
42	RAPH	Alain Gliksman	France	Andre Mauric	A.C.N. A.M.	Monohull
43	YAKSHA	Joan de Kat	France	Owner	Chantiers Naval de Limay	Trimaran
44	SIR THOMAS LIPTON	Geoffrey Williams	G.B.	Robert Clark	Derek Kelsall	Monohull
45	MAGUELONNE	Jean-Yves Terlain	France	Kim Holman	Henri Wauquiez	Monohull
47	AMBRIMA	Marco Cuiklinski	France	Andre Mauric	Yachting Service	Monohull

RIG	L.O.A. (Feet)	L.W.L. (Feet)	BEAM (Feet)	DRAFT (Feet)	SAIL AREA (Sq. ft.)	Displacement (Tons)	RESULT
Cutter	45·00	42·00	20·00	3·00	847	7·00	Mast broke
Sloop	35·00	25·50	9·70	6·00	532	6·90	12th
Sloop	37·50	27·20	10·70	6·20	705	6·10	14th
Sloop	34·75	32·00	9·50	6·00	536	6·50	9th
Sloop	29·25	21·70	8·80	4·60	321	2·45	Mast broke
Schooner	40·00	36·00	16·66	4·25	340	1·34	3rd
Ketch	57·40	42·60	13·40	8·20	1384	12·80	Rudder trouble
Sloop	49·20	39·30	30·00	2·00	645	2·50	Sank
Ketch	56·16	42·00	12·00	8·00	1200	12·00	1st
Sloop	34·60	25·00	9·10	5·80	488	6·50	10th
Sloop	35·40	25·90	8·86	4·90	441	3·90	Sank

Index